Alberto Salvio

Brane Worlds:

Theories with One or Two Extra Dimensions

MINKOWSKI
Institute Press

Alberto Salvio
Departamento de Física Teórica, Universidad Autónoma de Madrid
and Instituto de Física Teórica IFT-UAM/CSIC, Madrid, Spain
International School for Advanced Studies (SISSA/ISAS)
Via Beirut 2-4 34014, Trieste, Italy

ISBN: 978-1-927763-11-7 (softcover)
ISBN: 978-1-927763-12-4 (ebook)

Minkowski Institute Press
Montreal, Quebec, Canada
http://minkowskiinstitute.org/mip/

For information on all Minkowski Institute Press publications visit our
website at http://minkowskiinstitute.org/mip/books/

Preface

In this book we discuss some aspects concerning the construction of a 4D effective theory derived from a higher dimensional model. The first part is devoted to the study of how the heavy Kaluza-Klein modes contribute to the low energy dynamics of the light modes. The second part concerns the analysis of the spectrum arising from non standard compactifications of 6D minimal gauged supergravities, involving a warp factor and conical defects in the internal manifold.

To prepare the background for such topics, first we review standard Kaluza-Klein theories and brane world models.

Afterwards, in Part I, we introduce the study of the heavy mode contribution. We do so by discussing scalar models in arbitrary dimension and then by treating in some detail a 6D Einstein-Maxwell theory coupled to a charged scalar and fermions. The latter model has some interesting features as it can lead to a chiral low energy 4D effective theory, which is similar to the electroweak part of the Standard Model of Particle Physics. In this first part of the book our main interest is in the interaction terms. We point out that the contribution of the heavy KK modes is generally needed in order to reproduce the correct predictions for the observable quantities involving the light modes. In the 6D Einstein-Maxwell-Scalar model the contribution of the heavy KK modes are geometrically interpreted as the deformation of the internal space.

In Part II we introduce 6D minimal gauged supergravities, which are supersymmetric and non-Abelian extensions of the 6D model of Part I. We begin by summarizing the main features and possible applications of these models. Moreover we review warped brane solutions with 4D Poincaré invariance, and a compact and axisymmetric internal manifold, which, in a simple case, turns out to have conical defects. Afterwards we study fluctuations about these axisymmetric warped brane solutions. Much of our analysis is general and could be applied to other scenarios. We focus on bulk sectors that could give rise to Standard Model like gauge fields and charged matter. We reduce the dynamics to Schroedinger type equations plus physical boundary conditions, and obtain exact solutions for the Kaluza-Klein wave functions and discrete mass spectra. The power-law warping, as opposed to exponential in 5D, means that zero mode wave functions can be peaked on negative tension branes, but only at the price of localizing the whole Kaluza-Klein tower there. However, the codimension two defects allow the Kaluza-Klein mass gap to remain finite even in the infinite volume limit. In principle, in this scenario, not only gravity, but Standard Model fields could 'feel' the extent of large extra dimensions, and still be described by an effective 4D theory.

This book is mainly based on the author's Ph.D. thesis, which was completed during the fall of 2006. Since then a number of developments were done. Those related to Part I are summarized in Subsection 3.4, while extensions and applications of Part II are described in Subsection 5.5.

Contents

Introduction

The Standard Model (SM) of the electroweak and strong interactions, including neutrino masses, and the Einstein's theory of gravity successfully predict most of the physical quantities that we are able to measure. The former is the theoretical framework in which we usually study elementary particles physics and it consistently includes the principles of quantum mechanics and special relativity. The latter describes large scale structures such as the expanding universe and its history.

One of the main attractive features of the SM is the formulation of the non gravitational interactions as a consequence of a local invariance principle, generalizing the gauge invariance of electrodynamics. This leads to two additional fine structure constants associated to the weak and strong nuclear forces. In a similar way the Einstein's theory, or General Relativity (GR), includes gravitational interactions by requiring invariance under general space-time coordinate transformations.

Despite this similarity so far we do not have a complete and well understood quantum theory of gravity, which is valid at every energy scale. The most promising attempt to quantize GR is the Superstring Theory, which includes a graviton state in its physical spectrum. Moreover, Superstrings, having just one independent parameter, give hope to solve the theoretical problem of explaining the ad hoc structure of the gauge group and the quantum numbers of the SM. A consistent formulation of such theory requires a space-time dimensionality equal to 10 (or 11 in the case of M-Theory), necessarily leading to extra dimensions. The idea that we can live in a world with more than 4 dimensions was proposed by Kaluza and Klein before Superstring theory in order to unify gravitational and electromagnetic interactions. Indeed they analyzed a 5D Einstein's theory of gravity and proved that in the low energy limit this theory gives the 4D Einstein's theory and the Maxwell's theory of electrodynamics, if the extra dimension is *compactified*.

Now it is known that, remarkably, the presence of extra dimensions can help us to solve two longstanding problems of theoretical physics, which are related to the structure of the underlying quantum theory of gravity: the hierarchy problem and the cosmological constant problem. They both are fine-tuning problems because the former is related to the huge difference between the Planck scale and the electroweak scale, and the latter to the very small value of the observed vacuum's energy compared to the other energy scales (apart from neutrino masses).

Since we observe only 4 dimensions in our world, higher dimensional theories require a mechanism to hide extra dimensions. In the standard Kaluza-Klein (KK) theories the extra dimensions are invisible because they describe a smooth compact manifold with very small size, naturally of the order of the Planck length. More recently it has been proved that this is not actually necessary because SM fields can be localized on a (1+3)-dimensional sub-manifold, called 3-brane. However, every higher dimensional theory leads to an infinite number of particles from the 4D point of view. It is of course interesting to know whether or not it is possible to construct a 4D effective theory for the lightest particles, and, when this is possible, to study the role of the heavy particles in the low energy dynamics.

In the standard KK theories a higher dimensional field can be decomposed as an infinite but discrete sequence of 4D modes (KK modes), due to the compactness of the internal space. This sequence is called KK tower and the difference between two consecutive masses is of the order of the inverse proper radius of the internal space. However, in some cases a finite *mass gap* between the zero (or lightest) modes and the heavy KK modes can emerges even if the internal manifold is not compact. If this happens it seems clear that a low energy theory for the zero modes can be constructed by using the effective theory approach, namely *integrating out* the heavy degrees of freedom.

In a part of the present book we will study if the heavy KK modes can actually give a measurable contribution to the SM particles dynamics by assuming a finite mass gap. In particular we shall analyze the broken phase of the 4D effective theory, when the light modes acquire a vacuum expectation value (VEV), and the corresponding low energy mass spectrum. We will call this method the effective theory approach to spontaneous symmetry breaking (SSB). In order to study the heavy mode contribution to these observable quantities we shall also perform an alternative approach to SSB, in which we go directly to the broken phase by means of a solution of the higher dimensional EOM. We shall call the latter approach the *geometrical* approach to SSB. The final result will prove that in general the heavy mode contribution to the 4D effective theory is not negligible, because this is actually needed to reproduce the geometrical approach. Remarkably we will be able to prove that this is true even if the heavy modes are as massive as the Planck mass, contrary to the standard lore for which the physics at the Planck scale is not relevant for the dynamics at a lower scale, for instance at the electroweak scale.

In a second part of the present book we will consider a higher dimensional (in this case 6D) supergravity compactified on a manifold with singularities, which are not usually introduced in original KK theories. The specific model that we consider has interesting features in relation with both the hierarchy and the cosmological constant problem. Indeed recently Supersymmetric Large Extra Dimensions have been proposed as a possible scenario in which the cosmological constant can be *self-tuned* to the observed value if the dimension of the internal space is equal to 2. This is essentially due to a numerical coincidence between the inverse invariant radius of the internal space and the observed vacuum energy if the fundamental 6D Planck scale is at the electroweak scale. However, in order to implement this mechanism supersymmetry must be broken to allow a small but non vanishing cosmological constant. The class of solutions that we will consider are particularly interesting in this contest because they break supersymmetry completely. Such configurations present a *warp factor* and an internal manifold with 2 *conical singularities* and with the sphere topology. We shall study exactly and in great detail the KK towers for fermion and gauge field sectors that can contain SM fields. Moreover we will study the effect of conical singularities on such towers, finding that the mass gap is not necessarily equal to the inverse proper radius as in ordinary KK theories, with interesting application to Large Extra Dimensions.

The present book is organized as follows.

In Chapter 1 we will review standard KK theories and brane models by focusing on the details that are useful to understand the following chapters. In particular we will construct the 4D effective theory starting with the original 5D KK theory, we shall introduce the brane world idea by discussing the simple *kink* domain wall and finally we will summarize the main features of Large Extra Dimensions and Randall-Sundrum scenarios, which are both motivated by the hierarchy problem.

Afterwards we will present Part I of the book, which contains the results on the heavy mode contribution to the low energy dynamics. This part is divided in Chapters 2 and 3. In the former we will study the heavy mode contribution to the 4D effective theory in a simple scalar set up. This will allow us to explain in a rigorous way what we have proved and our steps. In Chapter 3 we will address the same problem but in a physically interesting contest. We will start with a 6D Einstein-Maxwell-Scalar model and consider the compactification over the sphere with a monopole background. This will lead to a chiral $SU(2) \times U(1)$ effective theory with an *Higgs field* triggering SSB from $SU(2) \times U(1)$ to $U(1)$. In this framework, which is similar to the electroweak part of the SM, we will show that the heavy KK modes, with masses of the order of the Planck scale, actually have effects on some measurable quantities of the low energy physics.

Part II is divided in Chapters 4 and 5. In the former we will review 6D supergravity focusing on the case in which we have the minimal number of supersymetries (minimal supergravity) and part of the R-symmetry group is gauged (gauged supergravity). Moreover we will review the Supersymmetric Large Extra Dimensions scenario and some singular solutions, which break completely supersymmetry. In Chapter 5 we will study gauge field and fermion fluctuations around non-smooth solutions. We will emphasize that these sectors can contain SM fields and we shall study the effect of (conical) singularities on the mass gap. Moreover we will analyze the particular case in which both the internal proper radius and the mass gap are large and the effect of this set up on the fundamental (6D) constants.

There are also three appendices. In Appendix A we give our conventions and notations, including recurrent abbreviations. In Appendix B we give the explicit computation of the 4D spectrum coming from the 6D Einstein-Maxwell-Scalar model. In Appendix C we provide some extra calculations concerning the 6D minimal gauged supergravity and we perform a stability analysis, considering all the present known anomaly free models of this type.

Chapter 1

Kaluza-Klein and Brane Models

In this chapter we review *KK theories* and *brane models* by focusing on the aspects that are relevant for Chapters 2, 3, and 5. The composition is as follows. In Section 1.1 we introduce KK theories by studying the simple example of 5D Einstein gravity compactified over S^1. In Section 1.2 we discuss the *kink* domain wall configuration that localizes matter fields on a *3-brane*. We will devote Section 1.3 to models with large extra dimensions, discussing in particular their relation with the hierarchy problem and their phenomenological implications. In Section 1.4 we will discuss 5D *warped* brane world models and the possibility of creating mass hierarchy in this contest. Furthermore, in Section 1.5 we review attempts that aim at a solution of the cosmological constant problem in theories with extra dimensions. Finally we will specifically consider 6D models in Section 1.6, where we will study *codimension 2* branes and their conical nature.

1.1 Kaluza-Klein Theories

The original motivation for studying field theories in space-time with more than 4 dimensions is to obtain a geometrical interpretation of internal quantum numbers such as the electric charge, that is to place them in the same context as energy and momentum [1, 2]. The latter observable quantities are associated with translational symmetry in

$(Minkowski)_4$, the 4D Minkowski space-time, whereas the internal observable quantities would be associated with symmetry motions in the extra dimensions.

In theories of the standard KK type[1] one assumes a D-dimensional $(D > 4)$ generally covariant field theory and, by some dynamical mechanism, obtains a partially compactified and *factorizable* background geometry,

$$M_4 \times K_d, \tag{1.1.1}$$

where M_4 is a 4D pseudo euclidean manifold, and K_d is a d-dimensional euclidean smooth compact manifold. The proper volume V_d of K_d must be sufficiently small to render the extra dimensions invisible. For instance for $V_d^{1/d} < 10^{-17} cm$, we expect the effects of K_d to be invisible up to energies of the order[2] of TeV. However, in order to support such compactification, extra matter fields in general are needed and therefore a completely geometrical explanation of the fundamental forces can be lost in this process. An interesting exception could be the anomaly free higher dimensional supergravities, in which supersymmetry and anomaly freedom can motivate the presence of additional matter fields.

The original works by Kaluza and Klein analysed a particular example of such a framework: the standard 5D Einstein-Hilbert theory compactified on $M_4 \times S^1$. The action of this model is

$$S = \frac{1}{\kappa^2} \int d^5 X \sqrt{-G} R, \tag{1.1.2}$$

where κ is a 5D Planck scale and other conventions are given in Appendix A. Compactification on S^1 means the physical equivalence

$$y \sim y + L, \tag{1.1.3}$$

where y is the fifth coordinate and L is the circumference of S^1. The five-dimensional metric separates into $G_{\mu\nu}$, $G_{\mu 5}$, and G_{55}. From the 4D point of view these are a metric, a vector, and a scalar. We can parametrize the metric as

$$ds^2 = G_{MN} dX^M dX^N = g_{\mu\nu} dx^\mu dx^\nu + G_{55} \left(dy + A_\mu dx^\mu\right)^2. \tag{1.1.4}$$

---

[1] For a review on this topic see [3].
[2] We use the following conversion relation: $(TeV)^{-1} = 10^{-17} cm$.

If $g_{\mu\nu}$, G_{55}, and A_μ depend on both x and y, (1.1.4) is the most general 5D metric, but henceforth we will assume they depend only on the noncompact coordinates x and, in this case, (1.1.4) is the most general metric invariant under x-dependent translations of y. That is, this form still allows the following reparametrizations

$$
\begin{aligned}
x^\mu &\rightarrow x'^\mu(x) \\
y &\rightarrow y + \Lambda(x),
\end{aligned} \tag{1.1.5}
$$

and under the latter

$$
A_\mu \rightarrow A_\mu - \partial_\mu \Lambda. \tag{1.1.6}
$$

So gauge transformations arise as part of the higher-dimensional coordinate group. This is the KK mechanism.

To see the effect of y-dependence, consider a massless complex scalar Φ in 5D. Relation (1.1.3) can be implemented by requiring Φ to be periodic with respect to y. Expanding the y-dependence of Φ in a complete set we have

$$
\Phi(X) = \sum_m \Phi_m(x) e^{i2\pi my/L}, \tag{1.1.7}
$$

where m is an integer. The momentum in the periodic dimension is quantized $p_y = 2\pi m/L$. The action for such a scalar is

$$
S_\Phi = -\int d^5 X \sqrt{-G} \partial_M \Phi^\dagger \partial^M \Phi. \tag{1.1.8}
$$

By using (1.1.7), and $\sqrt{-G} = \sqrt{-g} e^{\phi/2}$, where g is the determinant of $g_{\mu\nu}$ and $e^\phi = G_{55}$, we have

$$
\begin{aligned}
S_\Phi = -Le^{<\phi>/2} \sum_m \int d^4 x \sqrt{-g} \Big[\partial_\mu \Phi_m^\dagger \partial^\mu \Phi_m \\
+ \left(\frac{2\pi m}{Le^{<\phi>/2}} \right)^2 \Phi_m^\dagger \Phi_m \Big] + ...,
\end{aligned} \tag{1.1.9}
$$

where the dots represent interaction terms. Therefore Φ contains an infinite tower (KK tower) of 4D fields with squared mass

$$
M_m^2 = \left(\frac{2\pi m}{V_1} \right)^2, \tag{1.1.10}
$$

where $V_1 \equiv Le^{<\phi>/2}$ represents the invariant volume of the internal space S^1. We observe that the gap between two consecutive KK masses is fixed by the volume of the internal space. This is a general property of standard KK theories. Therefore, it seems that at energies small compared to V_1^{-1} only the zero modes $(M^2 = 0)$ can be physically relevant. However, integrating out the heavy modes in general gives a non-trivial contribution to the low energy dynamics. We will clarify this point in Chapters 2 and 3, which contains the work presented in [4].

The charge corresponding to the KK gauge invariance (1.1.6) is the p_y-momentum. In this simple example, all fields carrying the KK charge are massive. More generally there can be massless charged fields. We will provide an example in Chapter 3, where we will discuss a 6D Einstein-Maxwell-Scalar model compactified on $(Minkowski)_4 \times S^2$ with a non vanishing gauge field background. In this model the KK gauge group is $SU(2)$ and there are massless 4D fields in non-trivial $SU(2)$-representations.

We compute now the 4D effective action for the zero modes by putting the ansatz (1.1.4) in the Einstein-Hilbert term (1.1.2). The 5D Ricci scalar can be expressed in terms of the scalar field ϕ, the field strength $F_{\mu\nu}$ of A_μ, and the 4D metric $g_{\mu\nu}$:

$$R = R(g_{\mu\nu}) - 2e^{-\phi/2}\nabla^2 e^{\phi/2} - \frac{1}{4}e^\phi F_{\mu\nu}F^{\mu\nu}. \qquad (1.1.11)$$

Therefore the effective action for the zero modes is[3]

$$S_{eff} = \frac{L}{\kappa^2} \int d^4x \sqrt{-g}\, e^{\phi/2} \left(R(g_{\mu\nu}) - \frac{1}{4}e^\phi F_{\mu\nu}F^{\mu\nu} \right), \qquad (1.1.12)$$

where we have used the fact that $g_{\mu\nu}$, ϕ, and A_μ do not depend on y. We observe that the 4D Planck lenght[4] κ_4 is given by

$$\frac{1}{\kappa_4^2} = \frac{V_1}{\kappa^2}. \qquad (1.1.13)$$

On the other hand the effective gauge constant g_{eff} associated to

---

[3]The kinetic term for ϕ appears if one performs the Weyl transformation $g_{\mu\nu} \rightarrow e^{-\phi/2}g_{\mu\nu}$, which converts the gravitational term in the standard Einstein-Hilbert form.

[4]We define κ_4 in a way that the coefficient of $R(g_{\mu\nu})$ in the 4D lagrangian is $1/\kappa_4^2$.

$A'_\mu \equiv A_\mu/L$ is given by[5]

$$\frac{1}{g_{eff}^2} = \frac{V_1^2}{\kappa_4^2}. \tag{1.1.14}$$

Therefore the gauge constant is determined in terms of the 4D gravitational coupling and the volume of the internal space. If we require $g_{eff} \sim 1$, which is a natural choice as the SM gauge constants are of the order of 1, we get that the size of the internal space is naturally of the order of the 4D Planck length.

It is worth mentioning that this original KK theory can be generalized to include a smooth internal space of the form $\mathcal{G}/\mathcal{H}$ and develop an harmonic expansion, analogous to (1.1.7), on such coset space. In this way it is possible to show that a conventional Einstein-Yang-Mills model (realizing local $\mathcal{G}$-symmetry) emerges at the leading approximation [5]. Moreover it can be proved that this framework emerge as a 4D effective theory of a higher dimensional Einstein-Yang-Mills system [6].

Besides these attractive properties the original KK theory suffers from some phenomenological problems. It is not clear how to interpret the *radion* ϕ, because it represents a massless scalar particle that is not observed in nature. Moreover, it is not possible to get a 4D chiral fermionic spectrum in this framework, which is of course needed if one requires to reproduce the SM in the low energy limit.

The situation gets better if one includes bulk (D-dimensional) gauge fields [7] and considers generalizations of the original KK theory. A non-trivial example will be given in Chapter 3, based on a 6D Einstein-Maxwell-Scalar model, where the internal space will be taken to be $S^2 = SU(2)/U(1)$, and chiral fermions are obtained in the 4D effective theory.

Alternative solutions will be described in Sections 1.2, 1.3, and 1.4 where we will introduce the concept of *3-brane*. As we shall see the latter is a useful tool to address relevant problems of high energy physics like the hierarchy problem and the cosmological constant problem in higher dimensional models.

---

[5] We define g_{eff} in a way that the coefficient of $F'_{\mu\nu}F'^{\mu\nu}$ in the 4D lagrangian is $-1/(4g_{eff}^2)$.

1.2   Localized Wave Functions

The original KK idea assumes a compact internal space with a very small size to render the extra dimensions invisible. An interesting alternative can arise when this hypothesis is relaxed, but ordinary particles are confined inside a potential well, which is flat along the ordinary 4 dimensions and sufficiently steep along the d extra dimensions. As we shall see in this section the origin of such a potential can be purely dynamical, in the sense that it can emerge as a solution of the equations of motion (EOM). Therefore in this scenario we have spontaneous breaking of translation invariance. The ordinary matter can propagate in the D-dimensional space-time if it acquires high enough energy (basically if its energy exceeds the depth of the well).

1.2.1   The Domain Wall

The original idea of confining particles on a (1+3)-dimensional sub-manifold (*3-brane*) was proposed in [8] and independently in [9]. Now we illustrate the main idea by introducing the simplest higher dimensional model, which can give rise to matter field localization on a 3-brane (*brane world*). This is a 5D field theory with one real scalar living on $(Minkowski)_5$: the action is

$$S = \int d^5 X \left[-\frac{1}{2} \partial_M \varphi \partial^M \varphi - \lambda (\varphi^2 - v^2)^2 \right], \qquad (1.2.15)$$

where λ, and v are real parameters and we assume $\lambda \geq 0$ in order to have a bounded from below 5D potential. The internal symmetry of this theory is a Z_2 group: $\varphi \to \pm \varphi$.

   To derive the EOM from (1.2.15) through an action principle we require that the boundary terms in the integration by parts vanish. This leads to the conservation of current $J_M = \varphi \partial_M \varphi$, as explained in [10, 11, 12]:

$$\int d^5 X \partial_M \left(\varphi \partial^M \varphi \right) = 0. \qquad (1.2.16)$$

Actually we impose that for every pair of fields φ and φ' the condition $\int d^5 X \partial_M \left(\varphi \partial^M \varphi' \right) = 0$ is satisfied but in (1.2.16) the prime is understood. As usual we assume that the dependence on the 4D coordinates is such that Condition (1.2.16) reduces to

$$\left(\lim_{y \to +\infty} - \lim_{y \to -\infty}\right) \varphi \partial_y \varphi = 0, \qquad (1.2.17)$$

which involves only the dependence of φ on the extra dimension. A condition like (1.2.16) is usually used in brane world model to project out non physical modes[6]. Provided that (1.2.16) is satisfied, the EOM is

$$\partial_M \partial^M \varphi - 4\lambda(\varphi^2 - v^2)\varphi = 0, \qquad (1.2.18)$$

and we consider the following *kink* domain wall solution[7] [8]:

$$< \varphi > = v \tanh(\sqrt{2\lambda}vy) \equiv \varphi_c(y). \qquad (1.2.19)$$

The VEV in (1.2.19) spontaneously breaks the internal symmetry Z_2, and the translation invariance along the extra dimension.

To see how solution (1.2.19) produces a potential well along the extra dimension we consider perturbations around such a background solution. We define $\delta\varphi = \varphi - \varphi_c$ and the EOM at the linear level with respect to $\delta\varphi$ reads

$$\eta^{\mu\nu}\partial_\mu\partial_\nu\delta\varphi + \partial_y^2\delta\varphi - 4\lambda\left(3\varphi_c^2 - v^2\right)\delta\varphi = 0. \qquad (1.2.20)$$

The fluctuation $\delta\varphi$ must satisfy Condition (1.2.16) as well as φ itself. This will give us relevant information to construct the space of physical modes. We consider now a 4D plane wave solution of (1.2.20), that is we assume

$$\delta\varphi(x, y) = \mathcal{D}(y)e^{ik_\mu x^\mu} \qquad (1.2.21)$$

where k represents the 4D momentum of an ordinary particle with squared mass $M^2 = -k^2$. In this case the probability density along our 4D world is completely flat but we allow a non-trivial probability density $|\mathcal{D}|^2$ of finding the particle in an interval $[y, y + dy]$. Inserting (1.2.21) in Equation (1.2.20) we get

$$-\partial_y^2\mathcal{D} + 4\lambda\left(3\varphi_c^2 - v^2\right)\mathcal{D} = M^2\mathcal{D}. \qquad (1.2.22)$$

The latter equation is extensively studied in the literature [8, 13, 14,

---

[6]We will use analogous conditions in Chapter 5.
[7]For extended discussions on this solution see for example [13, 14, 15].

15]. It is a 1D Schroedinger equation with a potential

$$V(y) = 4\lambda v^2 [3\tanh^2(\sqrt{2\lambda}vy) - 1].\qquad(1.2.23)$$

This is exactly the potential well that we mentioned before. It is easy to see that the Condition (1.2.17) ensures that the hamiltonian $-\partial_y^2 + V(y)$ is hermitian with respect to the inner product $(\mathcal{D}', \mathcal{D}) = \int dy \mathcal{D}'(y)\mathcal{D}(y)$. To show that this potential effectively localizes particles on the brane we present the spectrum of the fluctuations. In the energy range $M \in [0, 2\sqrt{2\lambda}v]$ the spectrum is discrete and it consists of 2 states. There is a normalizable wave function with $M^2 = 0$:

$$\mathcal{D}_1(y) = \frac{N_1}{\cosh^2(\sqrt{2\lambda}vy)},\qquad(1.2.24)$$

where N_1 is a normalization constant. This wave function must represent the ground state for such a quantum mechanical problem because it does not have intersections with the y-axis. This automatically ensures that there are not tachyonic fluctuations. Moreover there is another normalizable wave function describing a bound state with mass $M = \sqrt{6\lambda}v$, as explained in [13]. This couple of normalizable states describes particles being confined inside the wall. For $M \geq 2\sqrt{2\lambda}v$ the spectrum is continous and represents perturbations that are not confined.

We will turn to the kink domain wall in Section 2.3 where we will introduce an additional scalar field whose lightest 4D mode will be interpreted as an Higgs field. We will study such model to provide a brane world example for the relevance of the heavy modes in the 4D low energy effective theory for the light modes, which is one of the main topics of the present book.

1.2.2   Fermion Zero Modes

In the simple model of Subsection 1.2.1 one can introduce fermions and study the corresponding localization problem, as for the $\delta\varphi$-fluctuations. If standard Yukawa couplings are present, a chiral low energy 4D fermion spectrum emerges, in the sense that only fermion zero modes with one chirality are localized on the wall.

To illustrate this point we introduce one 5D fermion with the following action [8]

$$S_F = \int d^5 X \left(\overline{\Psi} \Gamma^M \partial_M \Psi + h\varphi \overline{\Psi}\Psi \right), \qquad (1.2.25)$$

where h is a real constant and $\Gamma^\mu = \gamma^\mu$, $\Gamma^5 = \gamma^5$ and our conventions on γ^μ and γ^5 are given in Appendix A. The second term in (1.2.25) is a simple example of Yukawa coupling and it can provide a chiral localized zero mode. In order to show this, now we derive the EOM for Ψ. Similarly to the fluctuations coming from φ, which we have analysed in the previous section, in order to derive the EOM from (1.2.25) by means of an action principle we have to impose[8] the conservation of current $\overline{\Psi}\Gamma^M\Psi$ [16]:

$$\int d^5 X \partial_M \left(\overline{\Psi}\Gamma^M\Psi \right) = 0. \qquad (1.2.26)$$

Also in the fermion sector we assume that the dependence on the 4D coordinates is such that (1.2.26) reduces to

$$\left(\lim_{y \to +\infty} - \lim_{y \to -\infty} \right) \overline{\Psi}\gamma^5\Psi = 0. \qquad (1.2.27)$$

Thus the EOM reads

$$\Gamma^M \partial_M \Psi + h\varphi\Psi = 0. \qquad (1.2.28)$$

Now we decompose Ψ as follows: $\Psi = \Psi_R + \Psi_L$, where $\gamma^5 \Psi_R = \Psi_R$ and $\gamma^5 \Psi_L = -\Psi_L$, and therefore the EOM linearized around the kink background (1.2.19) are

$$\begin{aligned}
\gamma^\mu \partial_\mu \Psi_L + \partial_y \Psi_R + h\varphi_c \Psi_R &= 0, \\
\gamma^\mu \partial_\mu \Psi_R - \partial_y \Psi_L + h\varphi_c \Psi_L &= 0.
\end{aligned} \qquad (1.2.29)$$

In general this is a set of coupled differential equations, but in the case of zero modes ($\gamma^\mu \partial_\mu = 0$) they decouple. For instance Ψ_R satisfies

$$\partial_y \Psi_R = -h\varphi_c \Psi_R, \qquad (1.2.30)$$

---

[8]Actually we impose that for every pair of fields Ψ and Ψ' the condition $\int d^5 X \partial_M \left(\overline{\Psi}\Gamma^M\Psi' \right) = 0$ is satisfied but in (1.2.26) the prime is understood.

which can be easily solved:

$$\Psi_R(x, y) = \Psi_R^{(4)}(x)e^{-h\int_0^y dy' \varphi_c(y')}, \tag{1.2.31}$$

where $\Psi_R^{(4)}$ is a 4D chiral fermion, which does not depend on y. By using the explicit expression of φ_c given in (1.2.19), it is easy to see that (1.2.31) represents a bounded state, in the sense that its wave function profile along the extra dimension is peaked on $y = 0$ and rapidly goes to zero as $y \to \pm\infty$. On the other hand, the fermion with opposite chirality is unbounded, as can be seen by obtaining the corresponding wave function from (1.2.31) with the substitution $h \to -h$. The Condition (1.2.27) is automatically satisfied by these zero modes, but one chirality has to be projected out because it is not normalizable and therefore it has infinite kinetic energy from the 4D point of view. Therefore we conclude that the zero mode spectrum is effectively chiral as required by the SM.

These results, which we have obtained in a very simple model, can be generalized to include an arbitrary number of space-time dimensions and Yang-Mills and scalar backgrounds. In this more general framework, the conditions under which localized chiral fermions emerge are given in [17].

1.2.3   Gauge Field Localization

We conclude this section by examining gauge fields. Unlike the spinless and the spin-1/2 case, localizing gauge field wave functions is not simple, at least for massless non-Abelian fields. The reason is that we can have phenomenological problems when we construct the effective 4D gauge constants. Indeed, if we denote the zero mode wave function of a gauge field by $A_0(y)$ and the corresponding quantity for a fermion (or, in general, for a charged field) by $\Psi_0(y)$, usually the gauge constant in the 4D effective theory turns out to be proportional to an overlap integral of the form

$$\int dy \Psi_0^*(y) A_0(y) \Psi_0(y). \tag{1.2.32}$$

On the other hand, in the previous subsection we have seen that the fermion wave functions can be different for different types of particles as they depend on various parameters, for instance the constant h in (1.2.25). This can create a problem, as in non-Abelian gauge theories the gauge charges are quantized and fixed by group theory arguments (this is referred to as *charge universality*).

We can imagine some ways out. A first possibility is finding a theory that produces the same wave function profile for all the zero modes $\Psi_0(y)$. However, this is not the general case because, as we will see, in the explicit examples of Chapters 3 and 5 this does not happen. Therefore, a physical mechanism that ensures such an equality is needed, if one tries to solve the problem in this way.

A second possibility is obtaining a constant profile for the zero mode gauge fields, namely A_0 independent of y. Indeed, in this case the overlap integral defining the gauge constants becomes

$$A_0 \int dy \Psi_0^*(y) \Psi_0(y),$$

and it is proportional to the normalization constant of the fermion kinetic term in the 4D effective theory. Since we can normalize the fields in a way that

$$\int dy \Psi_0^*(y) \Psi_0(y) = 1, \qquad (1.2.33)$$

we obtain charge universality. An explicit example is the framework analyzed in Chapter 5, where the gauge field profiles are dynamically predicted to be constant.

Finally, in the literature there exists a more sophisticated mechanism to obtain localized gauge fields [18]. There it has been proposed to consider a gauge theory that is in *confinement phase* outside the brane, whereas it is in the Abelian Coulomb phase on the brane. This is achieved by means of a scalar field that acquires a kink VEV, such as φ in (1.2.19). Massless "quarks" and a U(1) gauge field are present on the brane and they cannot escape far away from the brane, since the lightest state of the confining theory of the bulk has a non-vanishing mass of the order[9] Λ. It is interesting to note that this mechanism provides both gauge field and fermion localization.

---

[9]The parameter Λ is the analogous of Λ_{QCD} in Quantum Chromodynamics.

1.3   Large Extra Dimensions

So far we have discussed KK theories, compactified on internal space with very small size, and brane world models with infinite extra dimensions, where a physical localization mechanism is needed to render the low energy physics effectively 4-dimensional. In this section we study an intermediate set up, in which the internal space is compact, like in original KK theories, but with *large size*. As we will show, this scenario, originally proposed by Arkani-Hamed, Dimopoulos and Dvali (ADD) [19], leads to an interesting reformulation of the hierarchy problem.

Contrary to what happen in original KK theories of Section 1.1, where the size of the internal space is naturally of the order of the Planck length, in the Large Extra Dimensions (LED) scenario the effects of extra dimensions are constrained by accessible experimental tests. In particular the behaviour of gravitational interactions should change at length scales below r, where

$$r \equiv (V_d)^{1/d}. \tag{1.3.34}$$

Moreover several possible collider experiments can detect extra dimensions if the KK mass gap (the mass gap between the zero modes and the first KK excited states) is of the order of

$$M_{GAP} \sim \frac{1}{r}, \tag{1.3.35}$$

as expected.

1.3.1   General Idea of ADD

To illustrate the main idea of ADD we consider an action containing the standard D-dimensional Einstein-Hilbert term[10]

$$S_{EH} = \frac{1}{\kappa^2} \int d^D X \sqrt{-G} R, \tag{1.3.36}$$

and we make the following ansatz:

$$ds^2 = e^{A(y)} g_{\mu\nu}(x) dx^\mu dx^\nu + g_{mn}(y) dy^m dy^n. \tag{1.3.37}$$

---

[10]See Appendix A for conventions and notations.

Namely we neglect 4D vectors and scalars coming from the D-dimensional metric but we keep the 4D metric $g_{\mu\nu}(x)$. The latter represents the complete 4D dynamical metric, including the VEV and the fluctuations. If one requires 4D Poincaré invariance one has also to impose $< g_{\mu\nu} >= \eta_{\mu\nu}$, but this is not necessary for our argument. In Eq. (1.3.37) we allow the presence of a non-trivial *warp factor* $e^{A(y)}$ to make the argument more general. This will be useful in Chapter 5 when we will discuss 6D warped brane worlds. This kind of spacetime are called *non-factorizable geometries* because they cannot be interpreted as products of manifolds like (1.1.1), as in the standard KK compactifications.

An explicit calculation leads to

$$R = e^{-A}R(g_{\mu\nu}) + ..., \tag{1.3.38}$$

where $R(g_{\mu\nu})$ is the Ricci scalar computed with $g_{\mu\nu}$, and the dots represent extra terms containing the warp factor and the metric components g_{mn}. So we obtain a 4D effective action for the gravitational field $g_{\mu\nu}$ as follows

$$S_{EH} = \frac{1}{\kappa^2} \int d^D X \sqrt{-\bar{G}} e^{-A} R(g_{\mu\nu}) + ... = \frac{1}{\kappa_4^2} \int d^4 x \sqrt{-g} R(g_{\mu\nu}) + ..., \tag{1.3.39}$$

where g is the determinant of $g_{\mu\nu}$ and the 4D Planck length κ_4 is given by

$$\frac{1}{\kappa_4^2} = \frac{V_d}{\kappa^2}, \tag{1.3.40}$$

where

$$V_d = \int d^d y \sqrt{-\bar{G}} e^{-A}, \tag{1.3.41}$$

and

$$\bar{G}_{MN} dX^M dX^N = e^{A(y)} \eta_{\mu\nu} dx^\mu dx^\nu + g_{mn}(y) dy^m dy^n.$$

We observe that V_d is a d-dimensional volume, which reduces to the volume of the internal space in the unwarped case ($e^A = 1$). Indeed the latter is defined by

$$\tilde{V}_d = \int d^d y \sqrt{g_d}, \tag{1.3.42}$$

where g_d is the determinant of the metric g_{mn} of the internal space. In terms of mass scales $M = 1/\kappa^{2/(D-2)}$, and $M_{Pl} = 1/\kappa_4$, Eq. (1.3.40) takes an illuminating form:

$$\left(\frac{M_{Pl}}{M}\right)^2 = (Mr)^d. \qquad (1.3.43)$$

If r is large compared to the fundamental length M^{-1}, the 4D Planck mass is much larger than the fundamental gravity scale M. One may push this line of reasoning to extreme and suppose that the fundamental gravity scale is of the same order as the electroweak scale, $M \sim TeV$. Then the hierarchy between M_{Pl} and the electroweak scale is entirely due to the large value of r. This is an interesting reformulation of the hierarchy problem because now it becomes the problem of explaining why r is large. We observe that r can be large because the volume $\tilde{V}_d$ of the internal space is large or because of a non-trivial contribution of the warp factor. In this section we will assume that only the former contribution is active. We will discuss the effect of the warp factor in Section 1.4.

1.3.2   Phenomenological Implications

In the ADD scenario new physics should emerge in the gravitational sector when the length scale r is reached. However, it is hard to test gravity at very short distances because it is a much weaker interaction than all the other forces. Over large distances gravity is dominant, however, as one starts going to shorter distances, electromagnetic forces are dominant and completely overwhelm the gravitational forces. This is the reason why the Newton-law of gravitational interactions has only been tested down to about a fraction of a millimeter. Today the bound on the size of the extra dimensions is $r \leq 0.1mm$, if only gravity propagates in the extra dimensions [20]. On the other hand, assuming $M \sim TeV$, we can calculate from (1.3.43) the value of r as a function of d,

$$r = M^{-1}\left(\frac{M_{Pl}}{M}\right)^{2/d} \sim 10^{32/d}\, 10^{-17} cm, \qquad (1.3.44)$$

where we used $M_{Pl} \sim 10^{16} TeV$. For one extra dimensions one obtains unacceptably large value of r. The case $d = 2$ is particularly interesting because it corresponds to $r \sim 1mm$. By increasing the fundamental mass scale of less than one order of magnitude ($M \sim TeV, 10TeV$) one gets a value of r close to the present bound ($r \leq 0.1mm$). Therefore for $d = 2$, one can hope to detect effects of the extra dimensions but still have a model, which is not ruled out by experiments. If $d > 2$ the size of the extra dimensions is less than $10^{-6}cm$, which is unlikely to be tested directly via gravitational measurements any time soon. For $d = 6$ (full dimensionality of space-time, as suggested by superstring theory), one has $r \sim 10^{-12}cm$, which is still much larger than the electroweak scale, $TeV \sim 10^{-17}cm$. In order the extra dimensions to be so large one has to find a physical mechanism to make the SM matter and gauge fields effectively 4-dimensional. The most popular way is to localize such fields on a 3-brane, but a possible alternative is to relax relation (1.3.35) and increase M_{GAP} considering compactifications on non-smooth space. We will show that this is possible in Chapter 5, which contains part of the work of [21], even if in this case the tuning of ADD scenario ($r \gg M^{-1}$) becomes a tuning on the fundamental bulk parameters.

Another interesting phenomenological consequence of this scenario is that extra dimensions should start to show up in collider experiments at energies approaching the TeV scale. If we assume (1.3.35) and that the only interactions, which can propagate in the bulk, is the gravitational interactions, the most distinctive feature of this scenario is the possibility to emit gravitons into the bulk. This process has strong dependence on the center of mass energy of particles colliding on the brane and has large probability at energies comparable to the fundamental gravity scale. Indeed even though the coupling of every KK graviton is weak, the total emission rate of KK gravitons is large at energies approaching M due to large number of KK graviton states. These particles will not be detected, so the typical collider processes will involve missing energy. For example the cross section of production of a KK graviton in the process

$$e^+ e^- \rightarrow \gamma + P_T, \qquad (1.3.45)$$

which involves a transverse missing particle P_T, is of the order of α/M_{Pl}^2, so the total cross section is of order $\sigma \sim \alpha N(E)/M_{Pl}^2$, where E is the center of mass energy, and $N(E)$ is the number of species of KK gravitons with mass below E. By using (1.3.35) we expect

$$N(E) \sim (Er)^d, \qquad (1.3.46)$$

which is indeed correct for the simple compactification on a d-dimensional torus. Therefore the total cross section becomes

$$\sigma \sim \frac{\alpha}{E^2} \left(\frac{E}{M} \right)^{d+2}$$

, which rapidly increases with E, and becomes comparable with the electromagnetic cross section at $E \sim M$. Processes like (1.3.45) have been analyzed in detailed in Ref. [22]. It has been found that both a 1 TeV e+e- collider and the CERN LHC will be able to reliably and perturbatively probe the fundamental gravity scale up to several TeV, with the precise value depending on the number d of extra dimensions.

It is interesting to note that relation (1.3.46) is relaxed if the KK mass gap is not of the order r^{-1}, which is possible if one considers non-smooth compactification. This can lead to a less stringent bound on the KK modes production.

1.4   Randall-Sundrum Models

Until now we have considered general warped geometries without giving an explicit example in a specific brane world scenario and without discussing the role of the warp factor. In this section we shall describe the most popular example of warped brane world, originally proposed by Randall and Sundrum (RS) [23, 24]. The original RS model is a 5D gravitational model whose action is the sum of the standard Einstein-Hilbert action, a 5D cosmological constant term, and a *3-brane action*:

$$S = \int d^5 X \sqrt{-G} \left(\frac{1}{\kappa^2} R - \Lambda \right) - T \int d^4 x \sqrt{-g} - T' \int d^4 x \sqrt{-g'},$$
(1.4.47)

where Λ is the 5D cosmological constant, T and T' are the brane tensions (energy densities) of two branes placed at $y = 0$ and $y = \pi r_c$ respectively, and g and g' are the determinants of the metrics $g_{\mu\nu}$ and $g'_{\mu\nu}$ induced on the branes:

$$g_{\mu\nu}(x) = G_{\mu\nu}(x, 0), \quad g'_{\mu\nu}(x) = G_{\mu\nu}(x, \pi r_c).$$
(1.4.48)

We assume that the branes are located at the boundaries of the extra dimensions, that is $0 \leq y \leq \pi r_c$. One can interpret one brane as our 4D world and the other one as an additional 4D world.

We consider now the most general background metric compatible with 4D Poincaré invariance:

$$ds^2 = e^{A(y)}\eta_{\mu\nu}dx^\mu dx^\nu + dy^2. \qquad (1.4.49)$$

The explicit expression for the warp factor can be found by putting this metric ansatz in the EOM, which follows from (1.4.47). The explicit calculation is given in the original work [23], so here we only give the final result. The existence of 4D flat solution requires fine-tunings between Λ, T, and T'. Indeed these constants are related in terms of a single scale k,

$$T = -T' = \frac{12k}{\kappa^2}, \quad \Lambda = -\frac{12k^2}{\kappa^2}, \qquad (1.4.50)$$

and the warp factor is given by

$$e^{A(y)} = e^{-2k|y|}. \qquad (1.4.51)$$

This fine-tuning is analogous to fine-tuning of the cosmological constant to zero in conventional 4D gravity.

Given the background we can now compute the volume V_1 defined in (1.3.41). The result is

$$V_1 = \frac{1}{2k}\left(1 - e^{-2k\pi r_c}\right). \qquad (1.4.52)$$

In this case we can have an hierarchy between M and M_{Pl} but only for large value of r_c, that is for a large internal space volume. However, we observe that M_{Pl} remains finite when r_c goes to infinity. This is an important property of the RS model, in which it is possible to have the ordinary 4D gravity even if we take the non-compact limit $(r_c \to \infty)$ for the internal space. Moreover, although the exponential has very little effect in determining the Planck scale, it plays a crucial role in the determination of the observable masses. Indeed since the induced brane metrics are related by $g'_{\mu\nu} = e^{-2k\pi r_c}g_{\mu\nu}$, any mass parameter m' on the $y = \pi r_c$ brane will correspond to a mass $m \equiv e^{-kr_c\pi}m'$ when measured on the $y = 0$ brane [23]. If $e^{kr_c\pi}$ is of order 10^{16}, this mechanism produces TeV physical mass scales from fundamental mass paramenters of the order of Planck mass. Because this geometric factor is an exponential, we do not require very large hierarchies among the fundamental parameters.

In the RS scenario one can localize ad hoc SM fields on a brane and exploit the mechanism that we have discussed so far in order to "solve" the hierarchy problem. However, after the original RS work models were developed in which SM fields originate from the bulk[11] [25, 26]. These authors tried to implement a localization mechanism similar to the one explained in section 1.2 but it is difficult to obtain a phenomenologically viable effective theory, because the various bulk fields are not always localized on a brane. We shall perform a similar study of bulk fields but in the contest of 6D warped brane worlds in Chapter 5.

1.5   Addressing the Cosmological Constant Problem

In Sections 1.3 and 1.4 we have discussed higher dimensional models, which give hope to solve the hierarchy problem or can reformulate it in an interesting way. The purpose of this section is to perform a similar study, but concerning the cosmological constant problem. In theories with extra dimensions such a problem can be reformulated as a problem of why the vacuum energy density has (almost) no effect on the observable quantities predicted by the 4D effective theory, which is valid at the energy range of present experiments. In particular this reformulation sounds suggestive in brane worlds scenarios, as it implies that the vacuum energy density may affect the bulk geometry, and that this may occur in such a way that the metric induced on our brane is (almost) flat. Roughly speaking, it seems plausible that, in the case of non-factorizable geometry, the vacuum energy density can induce a non-trivial warp factor, while the 4D Poincaré invariance remains unbroken. This possibility may exist irrespectively of the brane world picture [27].

In RS models with one compact or non-compact extra dimension, the warped solution, that we discussed in Section 1.4, requires a fine-tuning between the bulk cosmological constant and the brane tensions, which is explicitly given in (1.4.50). From this point of view the cosmological constant problem is not solved in these models, as the standard 4D fine-tuning is replaced by a similar 5D one.

---

[11]Although the Higgs field should be confined to the brane in order not to lose the gauge hierarchy.

A different scenario with non-compact internal manifold and vanishing cosmological constant in the bulk has been proposed in [28, 29]. In this works, besides a standard D-dimensional Einstein-Hilbert action, an additional 4D Einstein-Hilbert term and a 4D cosmological constant Λ_b are introduced on a 3-brane, giving rise to the following gravitational action

$$S = \frac{1}{\kappa^2} \int d^D X \sqrt{-G} R + \int d^4 x \sqrt{-g} \left(\frac{1}{\kappa_b^2} R(g_{\mu\nu}) - \Lambda_b \right). \quad (1.5.53)$$

Here we have two different Planck scales, κ in the bulk and κ_b on the brane. It has also been argued [28] that the 4D Einstein-Hilbert term is a natural ingredient, as quantum corrections generate it anyway. In [28] the above model has been formulated in 5D, but generalizations involving more than one extra dimension, which are relevant for the cosmological constant problem, have been considered in [29].

In this scenario the graviton is a metastable state and the essential feature of the model is the large distance modification of gravity. More precicely, the graviton propagator manifests a modified behaviour at length scales larger than r_c, where

$$r_c = \frac{M_{Pl}}{M^2}, \quad (1.5.54)$$

and $M_{Pl} = 1/\kappa_b$ and $M = 1/\kappa^{2/(D-2)}$. As a consequence, gravity does not necessarily react to sources that are relevant at length scale of the order $r_c \sim H_0^{-1}$, where H_0 is the Hubble constant, as is the case for the cosmological constant on the brane. As explained in [29, 30], this property could lead to a solution of the cosmological constant problem.

Other frameworks, which could be relevant for the cosmological constant problem, are 6D models with 2 compact extra dimensions and, more generally, self tuning models. These frameworks will be briefly reviewed in the next section and in Chapter 4, where we will deal with supersymmetric and large extra dimensions.

1.6 Codimension 2 Branes

As we discussed in Section 1.3 the case $d = 2$ is particularly interesting in the ADD scenario. It corresponds indeed to the smallest value of d compatible with tests of gravity. Moreover for $d = 2$ the ADD scenario predicts a value of r close to the present bound given by such experiments and therefore it is a falsifiable set up. Furthermore models in 6 dimensions have attracted interest as possible frameworks in which the cosmological constant could be faced, in both the non supersymmetric [31, 32, 33] and the supersymmetric [34]-[39] case. One of the motivations for that is the numerical equality between r and $\rho^{-1/4}$, where $\rho \sim (10^{-3}eV)^4$ is the measured vacuum's energy density, in 6D ADD scenario.

In order to implement these ideas usually *3-brane sources* are introduced in the action, namely terms similar to the second and third term in the RS action given in (1.4.47). However, branes whose transverse space is 2D (codimension 2 brane) are qualitative different from codimension 1 branes, described in Section 1.4. One of the main features of 6D models is the possibility to find 3-branes solutions of the EOM with a geometry independent of the value of the brane tensions, at least outside the branes. This property is not shared by the codimension 1 RS branes because of the following reason: the 5D cosmological constant Λ in that case depends on the brane tensions because of constraint (1.4.50), and, on the other hand, Λ gives a non vanishing contribution to the 5D Ricci scalar both outside and on the branes.

The aim of the present section is to describe codimension 2 branes in a simple set up because they are usefull to understand the results of Chapter 5, in which we will deal with 6D supersymmetric models having this type of brane solutions.

We start with the following 6D action

$$S = \frac{1}{\kappa^2} \int d^6 X \sqrt{-G} R - T \int d^4 x \sqrt{-g}, \qquad (1.6.55)$$

that is the sum of the standard 6D Einstein-Hilbert action and a single brane source. Moreover we consider the following simple ansatz for the background metric

$$ds^2 = \eta_{\mu\nu} dx^\mu dx^\nu + h(r) \left(dr^2 + r^2 d\varphi^2 \right), \qquad (1.6.56)$$

that is we assume 4D Poincaré invariance, vanishing warp factor, axisymmetry of the internal 2D space but we allow a non-trivial curvature in order to take into account the backreaction of the geometry due to the brane source. This curvature will be a functional of $h(r)$. We assume that r and φ range from 0 to ∞ and from 0 to 2π respectively. So we interpret r as a radial coordinate and φ as an angular coordinate.

The EOM associated to action (1.6.55) are the Einstein equations in presence of a brane source:

$$\frac{\sqrt{-G}}{\kappa^2}\left(R^{MN} - \frac{1}{2}RG^{MN}\right) = -\frac{T}{2}\delta^{(2)}(y)\sqrt{-g}\,g^{\mu\nu}\delta_\mu^M\delta_\nu^N, \quad (1.6.57)$$

where $y = (r\cos\varphi, r\sin\varphi)$, $\delta^{(2)}(y)$ is the 2D Dirac δ-function and G_{MN} are now the metric components in the coordinates x and y. By using ansatz (1.6.56) the μ,ν components of the Einstein equations give

$$\frac{1}{\kappa^2}\sqrt{g_2}R = T\delta^{(2)}(y), \quad (1.6.58)$$

where g_2 is the determinant of g_{mn}. We observe that the Ricci scalar outside the brane vanishes and therefore is independent of the brane tension, this is an example of the possible independence of geometry and brane tensions in codimension 2 brane world. The m,n components are trivially satisfied as they read $R_{mn} - \frac{1}{2}Rg_{mn} = 0$ and the 2D metric g_{mn} satisfies $R_{mn} = K(r)g_{mn}$ for some function $K(r)$. On the other hand the Ricci scalar is given in terms of h by

$$R = -\frac{1}{h}\nabla_E^2\ln h, \quad (1.6.59)$$

where ∇_E^2 is the covariant Laplacian computed with the euclidean metric $ds_E^2 = dr^2 + r^2 d\varphi^2$. By putting (1.6.59) in (1.6.58) we get a second order differential equation for h, which is satisfied by[12] $h \propto r^{2\zeta}$ where ζ is given in terms of the brane tension by

$$\zeta = -\frac{1}{4\pi}\kappa^2 T. \quad (1.6.60)$$

The parameter ζ has an interesting geometrical meaning that can be understood by introducing the coordinate $\rho \equiv r^{\zeta+1}/(\zeta+1)$. Indeed the 2D metric in terms of this coordinate is $d\rho^2 + (\zeta+1)^2\rho^2 d\varphi^2$. Therefore the effect of a non vanishing value of ζ is the shift $\varphi \to (1+\zeta)\varphi$, that

---

[12]In order to prove that $h \propto r^{2\zeta}$ is a solution one can use the relation $\nabla_E^2\ln r = 2\pi\delta^{(2)}(y)$.

is it produces a *deficit angle* δ given by [40]

$$\frac{2\delta}{\kappa^2} = T. \tag{1.6.61}$$

The resulting geometry presents a *conical defect* (or *conical singularity*) at $r = 0$. Eq. (1.6.61) is very important because it establishes a relation between a geometrical property of the internal space and a physical property of the 3-brane. We will use this formula in Chapter 5, where we will study the 6D supersymmetric and gravitational models compactified on an internal space that will turn out to have conical singularities.

Part I: Heavy Mode Contribution from Extra Dimensions

In studying the low energy physics of the light modes of a (4+d)-dimensional theory the attention is usually paid only to the spectral aspects. After determining the quantum numbers of the light modes the nature and the form of the interaction terms are often assumed to be dictated by symmetry arguments. Such arguments fix the general form of all the renormalilzable terms and if the effective theory is supersymmetric certain relationship between the couplings can also be established by supersymmetry. The masses are derived from the bilinear part of the effective action and the role of the heavy modes in the actual values of the masses and the couplings of the effective theory for the light modes are seldom taken into account. It is, however, well known from the study of the GUT's in 4-dimensions that the heavy modes have an important role to play even at low energies [41]. This happens through their contributions to the couplings entering into the effective Lagrangians describing the low energy physics of the light modes. According to Wilsonian approach, in order to obtain an effective theory applicable in large distances, the heavy modes should be integrated out [42]. The processes of "integrating out" has the effect of modifying the couplings of the light modes or introducing additional terms, which are suppressed by inverse powers of the heavy masses, as proved[13] in [43].

The aim of Part I of the present book is to examine the role of

---

[13]If the gauge symmetry is not assumed, the decoupling theorem of [43] in general does not hold [44, 45, 46].

the heavy modes in the low energy description of a higher dimensional theory. To this end we shall basically perform two complementary calculations. The first one will start from a solution of a higher dimensional theory with a 4D Poincaré invariance and develop an action functional for the light modes of the effective 4D theory. This effective action generally has a local symmetry, which should be broken by Higgs mechanism. Our interest is in the spectrum of the broken theory. The procedure is essentially what is adopted in the effective description of higher dimensional theories including superstring and M-theory compactifications. In this construction the heavy KK modes are generally ignored simply by reasoning that their masses are of the order of the compactification mass and this can be as heavy as the Planck mass. Therefore they cannot affect the low energy physics of the light modes.

In the second approach, which we shall call the *geometrical approach*, we shall find a solution of the higher dimensional equations with the same symmetry group as the one of the broken phase of the effective 4D theory for the light modes. We shall then study the physics of the 4D light modes around this solution. The result for the effective 4D theory will turn out to be *different* from the first approach. Our aim is to show that the difference is precisely due to the fact that in constructing the effective theory along the lines of the first approach the contribution of the heavy KK modes have been ignored. Indeed it will be argued - and demonstrated by working out some explicit examples - that taking due care of the role of the heavy modes a complete equivalence is established between the two approaches.

Part I contains two chapters and they both present results from [4]. In Chapter 2 we motivate the discussion in a simple context. In Section 2.1 we shall work out a simple model of two coupled scalar fields in 4-dimensions, which will be generalized to a multiplet of scalar fields in arbitrary dimensions in Section 2.2. The examples in Sections 2.1 and 2.2 will clarify the relevance of the heavy modes in the low energy description of the light modes. In Section 2.3 we shall discuss a simple 5D domain wall model including two bulk scalars, one of which acquires a kink VEV. In this simple higher dimensional model we will include also interactions in our study of the broken effective theory. In Chapter 3 we shall study a higher dimensional (in this case six dimensional) theory of Einstein-Maxwell system [7] coupled to a charged scalar and eventually also to charged fermions. We will define explicitly this model in Section 3.1. Such a model can arise in the compactification of string or M-theory to lower dimensions. The system has enough number of adjustable parameters to allow us to go to various limits in order to establish the main point of the present part of the book. The result will of course confirm the above mentioned expectation that in order to obtain a correct 4-dimensional description of the physics of the light modes the contribution of the heavy modes should be duly taken into account[14]. The explicit calculations will be given in Sections 3.2 and 3.3 and summarized in Section 3.4. This example is particularly interesting because the first kind of solution will produce an effective 4D gauge theory with a $SU(2) \times U(1)$ symmetry which will be broken to $U(1)$ by a complex triplet of Higgs fields. The geometrical approach, on the other hand, will take us directly to the unbroken $U(1)$ phase by deforming a round sphere into an ellipsoid[15]. In the geometrical approach the W and the Z masses originate from the deformation of the internal space. In this sense the standard Higgs mechanism acquires a geometrical origin[16]. We elaborate a little more on this point in Section 3.4 which summarizes our results. Some technical aspects of various derivations have been detailed in Appendix B.

---

[14]Of course this does not prove that the heavy mode contribution never vanishes: for instance [47] proves the decoupling of the heavy modes in the $(Minkowski)_4 \times S^2$ compactification of the 6D chiral supergravity [48], which is basically the supersymmetric version of our 6D theory.

[15] This will correspond to the magnetic monopole charge of 2. A monopole charge of unity will produce a Higgs doublet of SU(2).

[16]It should be mentioned that all of our discussion is (semi-) classical. To include quantum and renormalization effects is beyond the scope of the present study.

Chapter 2

Higher Dimensional Scalar Models

The present chapter will clarify the role of the heavy modes in the low energy dynamics without introducing any complications due to gauge and gravitational interactions. We will provide a generalization to a more sophisticated context in Chapter 3.

2.1 A Simple 4D Theory

Let us consider a 4D theory, which contains two real scalar fields φ and χ and with the lagrangian

$$\mathcal{L} = -\frac{1}{2}\partial_\mu\varphi\partial^\mu\varphi - \frac{1}{2}\partial_\mu\chi\partial^\mu\chi - \frac{1}{2}m_\varphi^2\varphi^2 - \frac{1}{2}m^2\chi^2 - \frac{1}{4}\lambda_\varphi\varphi^4 - \frac{1}{4}\lambda_\chi\chi^4 - a\varphi^2\chi^2,$$

where m_φ^2, m^2, λ_φ, λ_χ and a are real parameters[1]. Here we have the symmetry:

$$Z_2 : \varphi \to \pm\varphi,$$
$$Z_2' : \chi \to \pm\chi. \tag{2.1.1}$$

This is a very particular example and of course we do not want to present any general result in this section, we just want to provide a

---

[1]Of course we consider only the values of these parameters such that the scalar potential is bounded from below.

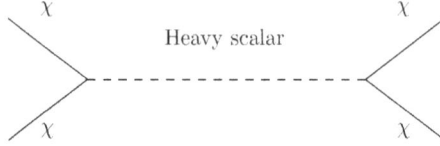

Figure 2.1: A tree diagram which describes the scattering of two light χ, through the exchange of an heavy scalar. This kind of diagram gives a contribution to the quartic term in the effective theory potential.

framework in which the general equivalence that we spoke about in the introduction emerges in a simple way and is not obscured by technical difficulties.

For $m_\varphi^2 < 0$ we have the following solution of the EOM:

$$\chi = 0, \qquad \varphi = \sqrt{\frac{-m_\varphi^2}{\lambda_\varphi}} \equiv \varphi_{eff}, \qquad (2.1.2)$$

which breaks Z_2 but preserves Z_2'. We can express the lagrangian in terms of the fluctuation $\delta\varphi$ and χ around this background:

$$
\begin{aligned}
\mathcal{L} = {} & -\frac{1}{2}\partial_\nu\delta\varphi\partial^\nu\delta\varphi - \frac{1}{2}\partial_\nu\chi\partial^\nu\chi + m_\varphi^2\,(\delta\varphi)^2 - \frac{1}{2}\mu^2\chi^2 \\
& -\sqrt{-m_\varphi^2\lambda_\varphi}\,(\delta\varphi)^3 - \frac{1}{4}\lambda_\varphi\,(\delta\varphi)^4 - \frac{1}{4}\lambda_{\chi\chi}\chi^4 - 2a\sqrt{\frac{-m_\varphi^2}{\lambda_\varphi}}\delta\varphi\chi^2 \\
& -a\,(\delta\varphi)^2\,\chi^2 + constants,
\end{aligned}
\qquad (2.1.3)
$$

where

$$\mu^2 \equiv m^2 - 2a\frac{m_\varphi^2}{\lambda_\varphi}. \qquad (2.1.4)$$

If $|\mu^2| \ll |m_\varphi^2|$, we expect that the heavy mode $\delta\varphi$ can be integrated out and an effective theory for χ can be constructed for both the signs of μ^2. However, it's important to note that $\delta\varphi$ cannot be simply neglected because it gives a contribution, because of the tri-linear[2] coupling $\delta\varphi\chi^2$ in (2.1.3), to the operator χ^4 in the effective theory, through the diagram 2.1. This is similar to what is usually done in GUT theories [41], where, for instance, four fermions effective interactions emerge by integrating out the heavy gauge fields [49].

---

[2] Also the quartic coupling $(\delta\varphi)^2\,\chi^2$ gives a contribution to the operator χ^4, but this is negligible in the classical limit.

At the classical level the effective lagrangian for χ is

$$\mathcal{L}_{eff} = -\frac{1}{2}\partial_\nu\chi\partial^\nu\chi - \frac{1}{2}\mu^2\chi^2 - \frac{1}{4}\left(\lambda_\chi - \frac{4a^2}{\lambda_\varphi}\right)\chi^4 + ...\,, \qquad (2.1.5)$$

where the dots represent higher dimensional operators. The term $a^2\chi^4/\lambda_\varphi$ is the contribution of the heavy mode. The result (2.1.5) was originally derived in [45], but here we want also to study the effective theory with spontaneous symmetry breaking and we want to compare it with the low energy limit of the fundamental theory.

For $\mu^2 > 0$, the minimum of the effective theory potential is for $\chi = 0$. Instead for $\mu^2 < 0$ we have

$$\chi = \sqrt{\frac{-\mu^2}{\lambda_\chi - \frac{4a^2}{\lambda_\varphi}}} \qquad (2.1.6)$$

and the fluctuation $\delta\chi$ over this background has the following mass squared:

$$M^2(\delta\chi) = -2\mu^2. \qquad (2.1.7)$$

This results will be not modified by the higher dimensional operator at the leading order[3] in μ. The equations (2.1.6) and (2.1.7) represent the effective theory prediction for the VEV and the spectrum in the phase where Z'_2 is broken.

On the other hand, a solution of the fundamental EOM, namely the EOM derived from the fundamental lagrangian $\mathcal{L}$, is

$$\begin{aligned}
\chi^2 &= \frac{-\mu^2}{\lambda_\chi - \frac{4a^2}{\lambda_\varphi}} + O(\mu^3), \\
\varphi^2 &= -\frac{m_\varphi^2}{\lambda_\varphi} + \frac{2a\mu^2}{\lambda_\varphi\lambda_\chi - 4a^2} + O(\mu^3)
\end{aligned} \qquad (2.1.8)$$

which is a small deformation of (2.1.2) at the leading non-trivial order in μ and breaks the Z'_2 symmetry. Moreover the light mode which corresponds to this solution has a mass squared $-2\mu^2$.

Therefore the effective theory prediction for the light mode VEV and spectrum is correct, at the order μ, in this simple framework, but the heavy mode contribution is necessary in order the effective theory prediction to be correct.

---

[3]The mass μ is small in the sense $|\mu| \ll |m_\varphi|$.

2.2   A More General Case

Now we want to extend the result of Section 2.1 and ref [45] to a more general class of theories. We consider a set of real D-dimensional scalars Φ_i with a general potential V: the lagrangian is

$$\mathcal{L} = -\frac{1}{2}\partial_M\Phi_i\partial^M\Phi_i - V(\Phi),\qquad(2.2.1)$$

where $M, N, ...$ run over all the space-time dimensions, while $\mu, \nu, ...$ and $m, n, ...$ are respectively the 4D and the internal coordinates indices. The EOM are

$$\partial_M\partial^M\Phi_i - \frac{\partial V}{\partial\Phi_i}(\Phi) = 0.\qquad(2.2.2)$$

We consider now a solution Φ_{eff} of (2.2.2) which preserves the 4D PoincarÃ© invariance and some internal symmetry group $\mathcal{G}$; the corresponding mass squared eigenvalue problem for the 4D states is

$$-\partial_m\partial_m\delta\Phi_i + \frac{\partial^2 V}{\partial\Phi_i\partial\Phi_j}(\Phi_{eff})\delta\Phi_j = M^2\delta\Phi_i,\qquad(2.2.3)$$

where $\delta\Phi$ is the fluctuation around Φ_{eff}. We assume that there are n normalizable solutions $\mathcal{D}_l$ with small eigenvalues ($M^2 \sim \mu^2$), other, in principle infinite, solutions[4] $\tilde{\mathcal{D}}_h$ with large eigenvalues ($M^2 \gg |\mu^2|$) and nothing else. These hypothesis are needed in order to define the concept of light KK modes.

We can expand the scalars Φ_i as follows

$$\Phi_i = (\Phi_{eff})_i + \chi_l(x)\mathcal{D}_{li}(y) + \tilde{\chi}_h(x)\tilde{\mathcal{D}}_{hi}(y),\qquad(2.2.4)$$

where χ_l and $\tilde{\chi}_h$ are respectively the light and heavy KK modes. We choose the $\mathcal{D}_l$ and $\tilde{\mathcal{D}}_h$ in order that they form an orthonormal basis for the functions over the internal space:

$$\langle\mathcal{D}_l|\mathcal{D}_{l'}\rangle \equiv \int d^{D-4}y\,\mathcal{D}_{li}(y)\mathcal{D}_{l'i}(y) = \delta_{ll'},$$

$$\left\langle\tilde{\mathcal{D}}_h|\tilde{\mathcal{D}}_{h'}\right\rangle \equiv \int d^{D-4}y\,\tilde{\mathcal{D}}_{hi}(y)\tilde{\mathcal{D}}_{h'i}(y) = \delta_{hh'},$$

$$\left\langle\mathcal{D}_l|\tilde{\mathcal{D}}_h\right\rangle \equiv \int d^{D-4}y\,\mathcal{D}_{li}(y)\tilde{\mathcal{D}}_{hi}(y) = 0.\qquad(2.2.5)$$

---

[4]In principle h can be a discrete or a continuous variable.

We note that χ_l and $\tilde{\chi}_h$ could both belong to some non-trivial representation of the internal symmetry group $\mathcal{G}$.

2.2.1 The Effective Theory Method

We construct now some relevant terms in the effective theory for the light KK modes χ_l. Here "relevant terms" mean relevant terms in the classical limit and in case we have a small point of minimum of the order μ of the effective theory potential: we want to compare the results of the effective theory for the light KK modes with the low energy limit of the fundamental theory expanded around a vacuum which is a small perturbation of Φ_{eff}. Further we calculate everything at leading non-trivial order[5] in μ. The relevant terms can be computed by putting just the light KK modes in the action and performing the integration over the extra dimensions and then by taking into account the effect of heavy KK modes through the diagrams like Fig. 2.1. In order to calculate those diagrams, we give the interactions between two light modes χ_l and one heavy mode $\tilde{\chi}_h$:

$$-\frac{1}{2}\left(\int d^{D-4}y V_{ijk}\mathcal{D}_{li}\mathcal{D}_{mj}\tilde{\mathcal{D}}_{hk}\right)\chi_l\chi_m\tilde{\chi}_h, \qquad (2.2.6)$$

where we have used the notation

$$V_{i_1...i_N} \equiv \frac{\partial^N V}{\partial\Phi_{i_1}...\partial\Phi_{i_N}}(\Phi_{eff}). \qquad (2.2.7)$$

We get the following relevant terms in the effective theory potential $\mathcal{U}$:

$$\mathcal{U}(\chi) = \frac{1}{2}c_l\mu^2\chi_l\chi_l + \frac{1}{3}\lambda^{(3)}_{lmp}\chi_l\chi_m\chi_p + \frac{1}{4}\lambda^{(4)}_{lmpq}\chi_l\chi_m\chi_p\chi_q + ..., \qquad (2.2.8)$$

where the dots represent non relevant terms, c_l are dimensionless numbers and

$$\lambda^{(3)}_{lmp} \equiv \frac{1}{2}\int d^{D-4}y V_{ijk}\mathcal{D}_{li}\mathcal{D}_{mj}\mathcal{D}_{pk}, \qquad (2.2.9)$$

$$\lambda^{(4)}_{lmpq} \equiv \frac{1}{3!}\left(\int d^{D-4}y V_{ijkk'}\mathcal{D}_{li}\mathcal{D}_{mj}\mathcal{D}_{pk}\mathcal{D}_{qk'}\right) + a_{lmpq}, \qquad (2.2.10)$$

---

[5]The μ mass scale is small in the sense $|\mu|$ is much smaller than the heavy masses.

where the quantities a_{lmpq} represent the heavy modes contribution and they are given by

$$a_{lmpq} = c_{lmpq} + c_{lpmq} + c_{lqpm} \qquad (2.2.11)$$

and

$$c_{lmpq} \equiv -\frac{1}{6} \int d^{D-4}y \, d^{D-4}y' V_{ijk}(y) \mathcal{D}_{li}(y) \mathcal{D}_{mj}(y)$$
$$\times G_{kk'}(y,y') V_{i'j'k'}(y') \mathcal{D}_{pi'}(y') \mathcal{D}_{qj'}(y'). \qquad (2.2.12)$$

The object $G_{kk'}$ is the Green function for the mass squared operator at the left hand side of (2.2.3) and it's explicitly given by

$$G_{kk'}(y,y') = \sum_h \frac{1}{m_h^2} \tilde{\mathcal{D}}_{hk}(y) \tilde{\mathcal{D}}_{hk'}(y'), \qquad (2.2.13)$$

where m_h^2 is the eigenvalue associated to the eigenfunction $\tilde{\mathcal{D}}_h$.

In the rest of this section we consider the predictions of the effective theory with spontaneous symmetry breaking. The potential (2.2.8) has to be considered as a generalization of (2.1.5), which was originally derived in [45]. A non vanishing VEV breaks in general $\mathcal{G}$ to some subgroup and it must satisfies

$$\frac{\partial \mathcal{U}}{\partial \chi_l} = c_l \mu^2 \chi_l + \lambda_{lmp}^{(3)} \chi_m \chi_p + \lambda_{lmpq}^{(4)} \chi_m \chi_p \chi_q = 0. \quad (2.2.14)$$

Since we require that χ_l goes to zero as μ goes to zero we have

$$\chi_l = \chi_{l1} + \chi_{l2} + \dots \qquad (2.2.15)$$

where χ_{l1} is proportional to μ, χ_{l2} is proportional to μ^2 and so on. At the order μ^2 the equations (2.2.14) reduce to

$$\lambda_{lmp}^{(3)} \chi_{m1} \chi_{p1} = 0 \qquad (2.2.16)$$

which implies

$$\lambda_{lmp}^{(3)} \chi_{p1} = 0. \qquad (2.2.17)$$

While, at the order μ^3, the equations (2.2.14) reduce to

$$c_l \mu^2 \chi_{l1} + \lambda_{lmpq}^{(4)} \chi_{m1} \chi_{p1} \chi_{q1} = 0, \qquad (2.2.18)$$

where we have used the equations (2.2.17).

Finally the mass spectrum corresponding to a solution of (2.2.14) is given by the eigenvalues of the hessian matrix of $\mathcal{U}$ in that solution:

$$\frac{\partial^2 \mathcal{U}}{\partial \chi_l \partial \chi_{l'}} = c_l \mu^2 \delta_{ll'} + 2\lambda^{(3)}_{ll'm} \chi_m + 3\lambda^{(4)}_{ll'mq} \chi_m \chi_q. \qquad (2.2.19)$$

If we assume, for simplicity, $\lambda^{(3)}_{ll'm} = 0$, which corresponds to the absence of cubic terms in $\mathcal{U}$, the leading order approximation of the hessian is simply given by

$$\frac{\partial^2 \mathcal{U}}{\partial \chi_l \partial \chi_{l'}} = c_l \mu^2 \delta_{ll'} + 3\lambda^{(4)}_{ll'mq} \chi_{m1} \chi_{q1} + O(\mu^3). \qquad (2.2.20)$$

In Subsection 2.2.2 we show that this matrix, which represents the mass spectrum for the light KK modes, and the equations (2.2.17) and (2.2.18) for the light modes VEVs are exactly reproduced by a D-dimensional analysis.

2.2.2   D-dimensional analysis

Now we present a D-dimensional (or geometrical) approach to compute low energy quantities: we want to find a solution of (2.2.2) which is a small perturbation, of the order μ, of Φ_{eff} and then we want to find the low energy mass spectrum of the fluctuations around this solution. In general this solution will break $\mathcal{G}$ to some subgroup like a solution of (2.2.14) does in the effective theory method. The explicit form of such solution in the simple case of Section 2.1 is given by (2.1.8) and the low energy mass spectrum in that simple case is represented by the squared mass $-2\mu^2$; now we want to generalize these results.

Let us consider the expansion (2.2.4); we observe that the statement that the solution is a small perturbation of Φ_{eff} means

$$\chi_l = \chi_{l1} + \chi_{l2} + \dots,$$
$$\tilde{\chi}_h = \tilde{\chi}_{h1} + \tilde{\chi}_{h2} + \dots, \qquad (2.2.21)$$

that is there are no big μ−independent terms in χ_l and $\tilde{\chi}_h$. We consider now a Taylor expansion of the equations (2.2.2) around Φ_{eff}:

$$\partial_m \partial_m \left(\Phi_i - (\Phi_{eff})_i \right)$$

$$-\sum_{k=1}^{N} \frac{1}{k!} V_{ii_1\ldots i_k} \left(\Phi_{i_1} - (\Phi_{eff})_{i_1} \right) \cdot \ldots \cdot \left(\Phi_{i_k} - (\Phi_{eff})_{i_k} \right)$$

$$+O(\mu^{N+1}) = 0. \tag{2.2.22}$$

At the order μ the equations (2.2.22) reduce to

$$(\partial_m \partial_m \delta_{ij} - V_{ij}) \left(\Phi_j - (\Phi_{eff})_j \right) + O(\mu^2) = 0, \tag{2.2.23}$$

which simply states

$$\tilde{\chi}_{h1} = 0. \tag{2.2.24}$$

Moreover at the order μ^2 the equations (2.2.22) imply

$$\tilde{\chi}_{h2} \left(\partial_m \partial_m \delta_{ij} - V_{ij} \right) \tilde{\mathcal{D}}_{hj} = \frac{1}{2} V_{ijk} \mathcal{D}_{lj} \mathcal{D}_{mk} \chi_{l1} \chi_{m1}, \tag{2.2.25}$$

which has two consequences: the first one is

$$\lambda_{lmp}^{(3)} \chi_{p1} = 0, \tag{2.2.26}$$

which can be derived from (2.2.25) by projecting over $\mathcal{D}_l$ and it exactly reproduces (2.2.17) of the effective theory method; the second consequence is

$$\tilde{\chi}_{h2} \tilde{\mathcal{D}}_{hi'}(y) = -\frac{1}{2} \chi_{l1} \chi_{m1} \int d^{D-4} y' G_{i'i}(y, y') V_{ijk}(y') \mathcal{D}_{lj}(y') \mathcal{D}_{mk}(y'), \tag{2.2.27}$$

where G still represents the Green function for the operator at the left hand side of (2.2.3). Now we can write the μ^3 part of the Eq. (2.2.22) as follows

$$-c_l \mu^2 \chi_{l1} \mathcal{D}_{li} - m_h^2 \tilde{\chi}_{h3} \tilde{\mathcal{D}}_{hi}$$

$$-\frac{1}{2} V_{ijk} \chi_{l1} \mathcal{D}_{lj} \left(\tilde{\chi}_{h2} \tilde{\mathcal{D}}_{hk} + \chi_{m2} \mathcal{D}_{mk} \right)$$

$$-\frac{1}{2} V_{ijk} \left(\chi_{l2} \mathcal{D}_{lj} + \tilde{\chi}_{h2} \tilde{\mathcal{D}}_{hj} \right) \chi_{m1} \mathcal{D}_{mk}$$

$$-\frac{1}{3!} V_{ijkk'} \mathcal{D}_{lj} \mathcal{D}_{mk} \mathcal{D}_{pk'} \chi_{l1} \chi_{m1} \chi_{p1} = 0. \tag{2.2.28}$$

If one projects this equation over $\mathcal{D}_l$ and uses the equations (2.2.26) and (2.2.27) one gets exactly the equations (2.2.18). Therefore, at the order μ, all the solutions of (2.2.14) are reproduced by the D-dimensional analysis and viceversa. Moreover we observe that these light KK modes VEVs, predicted by the effective theory, constitute approximate solutions of the fundamental D-dimensional EOM at leading non-trivial order because of the Eq. (2.2.24), which states that the heavy KK modes VEVs are higher order quantity with respect to the light KK modes VEVs.

Now we consider the mass squared eigenvalue problem which corresponds to a solution Φ; moreover we assume for simplicity $\lambda_{lmp}^{(3)} = 0$, like in the effective theory method. This eigenvalue problem is

$$\mathcal{O}_{ij}\delta\Phi_j \equiv -\partial_m\partial_m\delta\Phi_i + \frac{\partial^2 V}{\partial\Phi_i\partial\Phi_j}(\Phi)\delta\Phi_j = M^2\delta\Phi_i, \qquad (2.2.29)$$

where $\delta\Phi_i$ represents the fluctuations of the scalars around the solution Φ. We observe now that the equation (2.2.29) can be considered a time-independent Schrodinger equation: $\mathcal{O}$ is the hamiltonian and M^2 the generic energy level. Moreover we can perform a Taylor expansion of $\mathcal{O}$ around $\mu = 0$:

$$\mathcal{O} = \mathcal{O}_0 + \mathcal{O}_1 + \mathcal{O}_2 + ... \qquad (2.2.30)$$

The operators $\mathcal{O}_1$ and $\mathcal{O}_2$ can be easily expressed just in terms of χ_{l1} and χ_{l2} by using (2.2.4), (2.2.21) and the constraints (2.2.27) and (2.2.24) which come from the EOM. From the perturbation theory of quantum mechanics we know that the leading value of the low energy mass spectrum is given by the eigenvalues of the following mass squared matrix:

$$M_{ll'}^2 \equiv A_{ll'} + B_{ll'}, \qquad (2.2.31)$$

where

$$A_{ll'} \equiv\, < \mathcal{D}_l|\mathcal{O}_2|\mathcal{D}_{l'} > \qquad (2.2.32)$$

and

$$B_{ll'} \equiv\, -\sum_h \frac{1}{m_h^2} < \mathcal{D}_l|\mathcal{O}_1|\tilde{\mathcal{D}}_h >< \tilde{\mathcal{D}}_h|\mathcal{O}_1|\mathcal{D}_{l'} > . \qquad (2.2.33)$$

If one express the matrices A and B in terms[6] of χ_{l1} , one finds exactly the corresponding result (2.2.20) predicted by the effective theory.

So we have two equivalent (at least at the leading non-trivial order in μ) approaches to study the breaking of $\mathcal{G}$: the spontaneous symmetry breaking in the 4D effective theory and the D-dimensional analysis. We stress that, like in the simple model of Section 2.1, also in this more general case the heavy KK mode contribution in the effective theory can't be neglected if one wants to reproduce the D-dimensional result, even at the classical level. In general this is true not only in scalar theories but also in theories which involve gauge and gravitational interactions, as we illustrate in Chapter 3.

2.3   A Domain Wall Example

Here we provide an explicit application of previous results in the case of brane world models, where we have a non-trivial heavy KK mode contribution to the 4D effective theory. Moreover we study the role of heavy modes in the interactions of the 4D effective theory: we analyze the cubic interaction of the *Higgs field* after SSB, which is reproduced by the geometrical approach as well as masses and VEV.

2.3.1   The Model

We consider a 5D model with two scalar fields φ and ϕ. The lagrangian is

$$\begin{aligned}
\mathcal{L} = & -\frac{1}{2}\partial_M\varphi\partial^M\varphi - \frac{1}{2}\partial_M\phi\partial^M\phi - \frac{1}{2}m^2\phi^2 \\
& -\frac{1}{4}\xi\phi^4 - \lambda(\varphi^2 - v^2)^2 - \frac{\alpha}{2}\varphi^2\phi^2,
\end{aligned} \tag{2.3.1}$$

where m^2, ξ, λ, v and α are real parameters[7]. This model reduces to the one of Section 1.2 for $\phi = 0$. The internal symmetry of this theory is $Z_2 \times Z_2'$, where

$$\begin{aligned}
Z_2 &: \varphi \to \pm\varphi, \\
Z_2' &: \phi \to \pm\phi.
\end{aligned} \tag{2.3.2}$$

---

[6]The dependence on χ_{l2} disappears because we assume $\lambda_{lmp}^{(3)} = 0$, as one can easily check.

[7]We consider only the values of these parameters which correspond a potential bounded from below.

The EOM are

$$\partial_M \partial^M \varphi - 4\lambda(\varphi^2 - v^2)\varphi - \alpha\phi^2\varphi = 0,$$
$$\partial_M \partial^M \phi - m^2\phi - \xi\phi^3 - \alpha\varphi^2\phi = 0. \qquad (2.3.3)$$

Like in Section 1.2 we consider the domain wall solution for φ, whereas ϕ is assumed to vanish at the background level:

$$\phi = 0, \qquad \varphi = v\tanh(\sqrt{2\lambda}vy) \equiv \varphi_c(y). \qquad (2.3.4)$$

The VEV in (2.3.4) preserves the internal symmetry Z_2'. The mass squared eigenvalue problem which corresponds to solution (2.3.4), is

$$\mathcal{O}^{(1)}\delta\varphi \equiv \qquad -\partial_y^2\delta\varphi + 4\lambda\left(3\varphi_c^2 - v^2\right)\delta\varphi = M_{\delta\varphi}^2\delta\varphi, \qquad (2.3.5)$$
$$\mathcal{O}^{(2)}\phi \equiv \qquad -\partial_y^2\phi + m^2\phi + \alpha\varphi_c^2\phi = M_\phi^2\phi, \qquad (2.3.6)$$

where $\delta\varphi$ is the fluctuation of φ around φ_c. Equations (2.3.5) and (2.3.6) are studied in the literature [13, 8, 14, 15]. They are Schroedinger equations with a potential[8] $V(y) = a\tanh^2(\sqrt{2\lambda}vy) + b$, where a and b are constants. Like in Section 1.2 we can derive boundary conditions of the form (1.2.16) for both φ and ϕ. Therefore we project out exponentially growing solutions of (2.3.5) and (2.3.6). There is a wave function $\mathcal{D}_1$ for $\delta\varphi$ with $M_{\delta\varphi}^2 = 0$:

$$\mathcal{D}_1(y) = \frac{N_1}{\cosh^2(\sqrt{2\lambda}vy)}, \qquad (2.3.7)$$

where N_1 is a normalization constant such that $\int dy(\mathcal{D}_1(y))^2 = 1$. Concerning ϕ, we find an eigenfunction

$$\mathcal{D}_2(y) = \frac{N_2}{\cosh^\sigma(\sqrt{2\lambda}vy)}, \qquad (2.3.8)$$

where N_2 ensures $\int dy(\mathcal{D}_2(y))^2 = 1$. and

$$\sigma \equiv \frac{1}{2}\left(\sqrt{1 + 2\frac{\alpha}{\lambda}} - 1\right) \qquad (2.3.9)$$

---

[8]We require $\alpha > 0$ in order to get a localized wave function from equation (2.3.6).

The mass which corresponds to $\mathcal{D}_2$ is very small if we choose

$$m^2 = -(1 + \eta)M_0^2 \qquad (2.3.10)$$

where

$$M_0^2 \equiv \frac{2\alpha v^2}{\sqrt{1 + 2\frac{\alpha}{\lambda}} + 1} \qquad (2.3.11)$$

and η is a very small dimensionless parameter. Henceforth we assume (2.3.10) so the mass squared

$$M_\phi^2 = -\eta M_0^2 \equiv \mu^2, \qquad (2.3.12)$$

associated to $\mathcal{D}_2$, is small. The wave functions $\mathcal{D}_1$ and $\mathcal{D}_2$ are the ground states of the Schroedinger equations (2.3.5) and (2.3.6) respectively, because they have no nodes. All the remaining solutions of (2.3.5) and (2.3.6) have $M_{\delta\varphi}^2$, $M_\phi^2 \gg |\mu^2|$. We can perform the following expansion

$$
\begin{aligned}
\varphi(x, y) &= \varphi_c(y) + \chi_1(x)\mathcal{D}_1(y) + \sum_h \tilde{\chi}_{h1}(x)\tilde{\mathcal{D}}_{h1}(y), \\
\phi(x, y) &= \chi_2(x)\mathcal{D}_2(y) + \sum_h \tilde{\chi}_{h2}(x)\tilde{\mathcal{D}}_{h2}(y), \qquad (2.3.13)
\end{aligned}
$$

where $\tilde{\mathcal{D}}_{h1}$ and $\tilde{\mathcal{D}}_{h2}$ are the solutions of (2.3.5) and (2.3.6) with $M_{\delta\varphi}^2$, $M_\phi^2 \gg |\mu^2|$ and χ_i and $\tilde{\chi}_{hi}$ are respectively the light modes and the heavy modes.

2.3.2   The 4D Effective Theory for the Light Modes

The relevant terms in the effective theory for the light modes can be computed with the general argument given in Section 2.2.1. By using the residual Z_2' symmetry and the fact that $\mathcal{D}_i$ are even functions of y while φ_c is odd we find that the effective theory potential has the following form:

$$\mathcal{U}(\chi) = \frac{1}{2}\mu^2\chi_2^2 + \frac{1}{4}\lambda_1\chi_1^4 + \frac{1}{4}\lambda_2\chi_2^4 + a\chi_1^2\chi_2^2 + ..., \qquad (2.3.14)$$

where the dots represent higher order operators. The semiclassical approximation of λ_1, λ_2 and a including the heavy mode contribution is given by

$$\lambda_1 = \lambda_{1L} + \lambda_{1H}, \tag{2.3.15}$$

$$\lambda_2 = \lambda_{2L} + \lambda_{2H}, \tag{2.3.16}$$

$$a = a_L + a_H, \tag{2.3.17}$$

where λ_{1L}, λ_{2L} and a_L represent the light mode contribution and they are explicitly given by

$$\lambda_{1L} = 4\lambda \int dy (\mathcal{D}_1(y))^4, \quad \lambda_{2L} = \xi \int dy (\mathcal{D}_2(y))^4,$$

$$a_L = \frac{\alpha}{2} \int dy (\mathcal{D}_1(y))^2 (\mathcal{D}_2(y))^2, \tag{2.3.18}$$

whereas λ_{1H} λ_{2H} and a_H represent the heavy mode contribution:

$$\lambda_{1H} = -\frac{1}{2}(4!\lambda)^2 \int dy\, dy' (\mathcal{D}_1(y))^2 \varphi_c(y) G_1(y, y') (\mathcal{D}_1(y'))^2 \varphi_c(y'),$$

$$\lambda_{2H} = -2\alpha^2 \int dy\, dy' (\mathcal{D}_2(y))^2 \varphi_c(y) G_1(y, y') (\mathcal{D}_2(y'))^2 \varphi_c(y'),$$

$$a_H = a_{H1} + a_{H2}$$

$$= -12\alpha\lambda \int dy dy' (\mathcal{D}_2(y))^2 \varphi_c(y) G_1(y, y') (\mathcal{D}_1(y'))^2 \varphi_c(y')$$

$$- 2\alpha^2 \int dy dy' \mathcal{D}_1(y) \mathcal{D}_2(y) \varphi_c(y) G_2(y, y') \mathcal{D}_1(y') \mathcal{D}_2(y') \varphi_c(y'), \tag{2.3.19}$$

where the integrals over y are from $-\infty$ to $+\infty$, G_i are defined by

$$G_i(y, y') \equiv \sum_h \frac{1}{m_{hi}^2} \tilde{D}_{hi}(y) \tilde{D}_{hi}(y') \tag{2.3.20}$$

and m_{hi} is the mass of $\tilde{\chi}_{hi}$. The functions in (2.3.20) are the Green functions for the operators $\mathcal{O}^{(i)}$ defined in equations[9] (2.3.5) and (2.3.6):

$$\mathcal{O}^{(i)} G_i(y, y') = \delta(y - y') - \mathcal{D}_i(y) \mathcal{D}_i(y'). \tag{2.3.21}$$

---

[9]More precisely G_2 is the Green function of $\mathcal{O}^{(2)}$ with $m^2 = -M_0^2$.

The general solution of (2.3.21) is

$$
\begin{aligned}
G_i(y,y') &= c_i^{(1)}(y')\mathcal{D}_i(y) + c_i^{(2)}(y')\mathcal{D}_i^{\perp}(y) \\
&+ \theta(y - y')\left(\mathcal{D}_i(y')\mathcal{D}_i^{\perp}(y) - \mathcal{D}_i(y)\mathcal{D}_i^{\perp}(y')\right) \\
&+ \mathcal{D}_i(y')\left[\mathcal{D}_i(y)\int_0^y dy'' \mathcal{D}_i(y'')\mathcal{D}_i^{\perp}(y'')\right. \\
&\left. - \mathcal{D}_i^{\perp}(y)\int_0^y dy''(\mathcal{D}_i(y''))^2\right],
\end{aligned}
\tag{2.3.22}
$$

where $c_i^{(1)}(y')$ and $c_i^{(2)}(y')$ are generic functions of y' and

$$
\mathcal{D}_i^{\perp}(y) \equiv -\mathcal{D}_i(y)\int_0^y \frac{dy'}{(\mathcal{D}_i(y'))^2};
$$

$\mathcal{D}_i$ and $\mathcal{D}_i^{\perp}$ are two independent eigenfunctions of $\mathcal{O}^{(i)}$ with vanishing eigenvalue. In our case we can compute $c_i^{(2)}(y')$ by requiring $G_i(y,y')$, as function of y, to have no exponentially growing part: we get

$$
c_i^{(2)}(y') = -\frac{1}{2}\mathcal{D}_i(y').
\tag{2.3.23}
$$

Moreover, from (2.3.20), we are forced to project $G_i(y,y')$ in the subspace orthogonal to $\mathcal{D}_i(y)$, this fixes $c_i^{(1)}(y')$; however, here we do not need the explicit expression for $c_i^{(1)}(y')$.

Computation of the Effective Coupling Constants

By using the explicit form of $\mathcal{D}_i$ and G_i we can compute λ_1, λ_2 and a by means of equations (2.3.18) and (2.3.19). The expressions of such coupling constants simplify because $\mathcal{D}_i$ is even and φ_c is odd under $y \to -y$; in particular first and third line in (2.3.22) do not give any contribution to λ_1, λ_2 and a because of those parities. Concerning λ_1, a non-trivial balancing between the heavy mode contribution and the light modes one gives

$$
\lambda_1 = 0.
\tag{2.3.24}
$$

The explicit expression for λ_2 is

$$
\lambda_2 = \frac{N_2^4}{v\sqrt{2\lambda}}\xi J_L(\sigma)\left(1 + \frac{2a^2}{\lambda\xi}\frac{J_H(\sigma)}{J_L(\sigma)}\right),
\tag{2.3.25}
$$

where
$$J_L(\sigma) \equiv \int_{-\infty}^{+\infty} \frac{dx}{\cosh^{4\sigma}(x)} = \frac{\sqrt{\pi}\,\Gamma(2\sigma)}{\Gamma(\frac{1}{2} + 2\sigma)}, \tag{2.3.26}$$

where Γ is the Euler gamma function, and

$$J_H(\sigma) = -\frac{1}{2 + 2\sigma} \int_{-\infty}^{+\infty} dx \frac{\tanh(x)}{\cosh^{4+4\sigma}(x)}$$
$$\times \left(\frac{3}{8}x + \frac{1}{4}\sinh(2x) + \frac{1}{32}\sinh(4x) \right). \tag{2.3.27}$$

The ratio
$$\rho \equiv \frac{2\alpha^2}{\lambda\xi} \frac{J_H(\sigma)}{J_L(\sigma)} \tag{2.3.28}$$

in (2.3.25) represents the heavy mode contribution to this coupling constant. In general ρ is not negligible: for instance, by choosing $\sigma = 2$ that is $\alpha = 12\lambda$, we get

$$\rho = -\frac{1}{3}\frac{\alpha}{\xi}, \tag{2.3.29}$$

which is not small for $\alpha, \xi \sim 1$. Finally we note that a depends on μ^2, because of G_2 in a_{H2}. We perform a Taylor expansion of this coupling around $\mu^2 = 0$:
$$a = a_0 + O(\mu^2), \tag{2.3.30}$$

where

$$a_0 = N_1^2 N_2^2 \frac{\alpha}{v\sqrt{2\lambda}} \left(\frac{1}{2} I_L(\sigma) + 6I_{H1}(\sigma) + \frac{\alpha}{\lambda} I_{H2}(\sigma) \right), \tag{2.3.31}$$

and

$$I_L(\sigma) \equiv \int_{-\infty}^{+\infty} \frac{dx}{\cosh^{2\sigma+4}(x)},$$

$$I_{H1}(\sigma) \equiv \int_{-\infty}^{+\infty} dx \int_{-\infty}^{+\infty} dx' \frac{\tanh(x)}{\cosh^{2+2\sigma}(x)} \theta(x-x')$$
$$\times \int_{x'}^{x} ds \cosh^4(s) \frac{\tanh(x')}{\cosh^6(x')},$$

$$I_{H2}(\sigma) \equiv \int_{-\infty}^{+\infty} dx \int_{-\infty}^{+\infty} dx' \frac{\tanh(x)}{\cosh^{2+2\sigma}(x)} \theta(x-x')$$
$$\times \int_{x'}^{x} ds \cosh^{2\sigma}(s) \frac{\tanh(x')}{\cosh^{2+2\sigma}(x')},$$

If we set for simplicity $\sigma = 2$, we get $a_0 = 0$ again by means of a non-trivial balance between light and heavy mode contribution.

Broken Effective Theory and the Role of Heavy Modes

Here we show that the heavy modes have a non-trivial role in the low energy physics. The potential now looks like

$$\mathcal{U}(\chi) = \frac{1}{2}\mu^2\chi_2^2 + \frac{1}{4}\lambda_2\chi_2^4 + a\chi_1^2\chi_2^2 + \dots, \tag{2.3.32}$$

where the dots represent higher order terms (powers of $\chi_{1,2}$ greater than 4) and $a = O(\mu^2)$.

A consistent vacuum at the leading order in μ is (for $\mu^2 < 0$):

$$< \chi_1 > = 0, \quad < \chi_2 > = \sqrt{\frac{-\mu^2}{\lambda_2}} \tag{2.3.33}$$

Vacuum (2.3.33) spontaneously breaks Z_2'. The corresponding leading order mass spectrum of the fluctuations around (2.3.33) is

$$M_1^2 = 0, \quad M_2^2 = -2\mu^2. \tag{2.3.34}$$

We observe that the contribution of the heavy modes to $< \chi_2 >$ is not trivial because the quantity ρ in (2.3.29) is not negligible. This contribution is a modification of the cubic self-interaction of $H \equiv$

$\chi_2 - <\chi_2>$:

$$\mathcal{U}(\chi_1, H) = \lambda_2 < \chi_2 > H^3 + O(\mu^2). \qquad (2.3.35)$$

We observe that this is the only cubic interaction at the order μ. Such an interaction is reproduced by the 5D analysis explained in Subsection 2.2.2, as one can expect, only if the heavy mode contributions are taken into account, that is only if the quantity ρ in (2.3.28) is not neglected.

So in this section we have extended the results of Sections 2.1 and 2.2, concerning only the spectral aspects of the broken effective theory, by including the description of the heavy mode contribution to interactions in the broken effective theory. Also at the interaction level such contribution turns out to be needed to reproduce the D-dimensional, in this case 5-dimensional, approach.

However, we observe that the models presented in this chapter do not include gauge and gravitational interactions. Therefore in the next chapter we will introduce a gauge and gravitational model and we will discuss again the role of the heavy modes in the effective theory taking into account the extra terms in the lagrangian that are implied by gravity and gauge invariance.

Chapter 3

6D Einstein-Maxwell-Scalar Model

In Chapter 2 we have proved that the heavy mode contribution is necessary to reproduce the correct low energy dynamics because, without this contribution, the 4D effective theory approach cannot reproduce in general the D-dimensional (or geometrical) approach to spontaneous symmetry breaking. The aim of the present chapter is to prove a similar statement in a more interesting context which can be extended to a semi-realistic theory. Our model will be an Einstein-Maxwell-Scalar model in 6D, which will be compactified over an internal space with the S^2 topology. This is an ordinary KK theory in which the KK mass scale is naturally of the order of the Planck mass. We will be able to prove that the heavy mode contribution is not negligible even if they have such a large mass.

3.1 Definition of the Model and 6D Equations of Motion

We consider a 6D field theory of gravity with a $U(1)$ gauge invariance, including a charged scalar field ϕ and eventually fermions. The bosonic

49

action is[1]

$$S_B = \int d^6 X \sqrt{-G} \left[\frac{1}{\kappa^2} R - \frac{1}{4} F_{MN} F^{MN} - (\nabla_M \phi)^* \nabla^M \phi - V(\phi) \right],$$
(3.1.1)

where R is the Ricci scalar, κ represents the 6D Planck scale, F_{MN} is the field strength of the $U(1)$ gauge field A_M, defined by

$$F_{MN} = \partial_M A_N - \partial_N A_M$$
(3.1.2)

and

$$\nabla_M \phi = \partial_M \phi + ie A_M \phi,$$
(3.1.3)

where e is the $U(1)$ gauge coupling. Moreover V is a scalar potential and we choose

$$V(\phi) = m^2 \phi^* \phi + \xi (\phi^* \phi)^2 + \lambda,$$
(3.1.4)

where m^2 and ξ are generical real constants, with the constraint $\xi > 0$ and λ represents the 6D cosmological constant.

From the action (3.1.1) we can derive the general bosonic EOM. However, we focus on the following class of backgrounds, which are invariant under the 4D Poincaré group:

$$
\begin{aligned}
ds^2 &= \eta_{\mu\nu} dx^\mu dx^\nu + g_{mn}(y) dy^m dy^n. & (3.1.5) \\
A &= A_m(y) dy^m, & (3.1.6) \\
\phi &= \phi(y), & (3.1.7)
\end{aligned}
$$

where g_{mn} is the metric of a 2-dimensional compact internal manifold K_2; so the 6D space-time manifold is $(Minkowski)_4 \times K_2$. By using (3.1.5), (3.1.6) and (3.1.7), we can write the bosonic EOM in the following form:

$$
\begin{aligned}
&\nabla^2 \phi - m^2 \phi - 2\xi (\phi^* \phi) \phi = 0, \\
&\nabla_m F^{mn} + ie \left[\phi^* \nabla^n \phi - (\nabla^n \phi)^* \phi \right] = 0, \\
&\frac{1}{\kappa^2} R_{mn} - \frac{1}{2} F_{mp} F_n{}^p - \frac{1}{2} (\nabla_m \phi)^* \nabla_n \phi - \frac{1}{2} (\nabla_n \phi)^* \nabla_m \phi = 0, \\
&\frac{1}{4} F^2 - \lambda - m^2 \phi^* \phi - \xi (\phi^* \phi)^2 = 0,
\end{aligned}
$$
(3.1.8)

where $\nabla^2 \equiv \nabla_m \nabla^m$ is the covariant Laplacian over the internal man-

---

[1]Our conventions are fixed in Appendix A.

ifold. The equations (3.1.8) must be satisfied by the bosonic VEV.

We introduce also fermions and gauge invariant coupling with the scalar ϕ. In order to do that it is necessary to introduce at least a pair of 6D *Weyl spinors* ψ_+ and ψ_-, where ψ_+ and ψ_- are eigenvectors of Γ^7 with eigenvalues $+1$ and -1 respectively[2]. We consider the following fermionic action:

$$S_F = \int d^6 X \sqrt{-G} (\overline{\psi_+} \Gamma^M \nabla_M \psi_+ + \overline{\psi_-} \Gamma^M \nabla_M \psi_-$$
$$+ g_Y \phi^* \overline{\psi_+} \psi_- + g_Y \phi \overline{\psi_-} \psi_+), \qquad (3.1.9)$$

where g_Y is a real Yukawa coupling constant. In (3.1.9) ∇_M represents the covariant derivative acting on spinor, which includes the gauge and the spin connection. The $U(1)$ charge e_+ and e_- of ψ_+ and ψ_- have to satisfy the condition $e_- = e_+ + e$ coming from the gauge invariance of the Yukawa terms. In the following we consider the choice $e_+ = e/2$ and $e_- = 3e/2$, corresponding to a simple harmonic expansion for the compactification over $(Minkowski)_4 \times S^2$. From (3.1.9) we get the following EOM:

$$\Gamma^M \nabla_M \psi_+ + g_Y \phi^* \psi_- = 0, \quad \Gamma^M \nabla_M \psi_- + g_Y \phi \psi_+ = 0. \qquad (3.1.10)$$

Now we define the following 4D Weyl spinors:

$$\psi_{\pm L} = \frac{1 - \gamma^5}{2} \psi_\pm, \quad \psi_{\pm R} = \frac{1 + \gamma^5}{2} \psi_\pm, \qquad (3.1.11)$$

where γ^5 is the 4D chirality matrix. In terms of $\psi_{\pm L}$ and $\psi_{\pm R}$ the EOM, for a $(Minkowski)_4 \times K_2$ background space-time, are[3]

$$\left(\partial^2 + 2\nabla_+ \nabla_- - g_Y^2 |\phi|^2 \right) \psi_{+L} - \sqrt{2} g_Y \left(\nabla_+ \phi^* \right) \psi_{-L} = 0,$$
$$\left(\partial^2 + 2\nabla_- \nabla_+ - g_Y^2 |\phi|^2 \right) \psi_{-L} - \sqrt{2} g_Y \left(\nabla_- \phi \right) \psi_{+L} = 0,$$
$$\left(\partial^2 + 2\nabla_- \nabla_+ - g_Y^2 |\phi|^2 \right) \psi_{+R} + \sqrt{2} g_Y \left(\nabla_- \phi^* \right) \psi_{-R} = 0,$$
$$\left(\partial^2 + 2\nabla_+ \nabla_- - g_Y^2 |\phi|^2 \right) \psi_{-R} + \sqrt{2} g_Y \left(\nabla_+ \phi \right) \psi_{+R} = 0, \qquad (3.1.12)$$

---

[2] Our conventions for the 6D gamma matrices are given in Appendix A.

[3] We rearrange the equations in a way that the left handed and right handed sector are split.

where $\partial^2 \equiv \eta^{\mu\nu}\partial_\mu\partial_\nu$,

$$\nabla_\pm = \frac{1}{\sqrt{2}}(\nabla_5 \pm i\nabla_6) \tag{3.1.13}$$

and $\nabla_{5,6}$ are the covariant derivative components in an orthonormal basis. The equations (3.1.12) will be used in order to compute the fermionic spectrum.

3.2   4D Electroweak Symmetry Breaking

3.2.1   The $SU(2) \times U(1)$ Background Solution

An $SU(2) \times U(1)$-invariant solution of (3.1.8) is [7]

$$
\begin{aligned}
ds^2 &= \eta_{\mu\nu}dx^\mu dx^\nu + a^2\left(d\theta^2 + \sin^2\theta d\varphi^2\right), & (3.2.1)\\
A &= \frac{n}{2e}(\cos\theta - 1)d\varphi \equiv -\frac{n}{2e}e^3(y), & (3.2.2)\\
\phi &= 0, & (3.2.3)
\end{aligned}
$$

subject to the constraints

$$\lambda = \frac{n^2}{8e^2 a^4} = \frac{1}{\kappa^2 a^2}, \tag{3.2.4}$$

where n is the monopole number. The metric (3.2.1) is the sum of the 4D Minkowski metric and the metric of the 2D sphere S^2, with radius a. So we have $K_2 = S^2$ and our internal space is maximally symmetric. We use the spherical coordinates θ and φ, so $dy^5 = a\,d\theta$, $dy^6 = a\,d\phi$. The 1-form (3.2.2) is a monopole configuration for the $U(1)$ gauge field. In (3.2.2) A is expressed in the chart $0 \le \theta < \pi$, $0 \le \varphi < 2\pi$. Instead in the chart $0 < \theta \le \pi$, $0 \le \varphi < 2\pi$, A has the form

$$A = \frac{n}{2e}(\cos\theta + 1)d\varphi. \tag{3.2.5}$$

The two 1-forms (3.2.2) and (3.2.5) must differ by a single valued gauge transformation and so we have that n is an integer. This rule is called Dirac quantization condition. We note that the solution in (3.2.1), (3.2.2) and (3.2.3) has an $SU(2) \times U(1)$ symmetry. It's useful to introduce an orthonormal basis in the internal cotangent space [7].

We choose the following 1-forms basis

$$e^{\pm}(y) = \pm \frac{i}{\sqrt{2}} e^{\pm i\varphi} \left(d\theta \pm i\sin\theta d\varphi \right).$$
(3.2.6)

In this basis the metric (3.2.1) has the form

$$ds^2 = \eta_{\mu\nu} dx^\mu dx^\nu + a^2 \left(e^+ e^- + e^- e^+ \right).$$
(3.2.7)

Under a rotation on the sphere we have [7]

$$e^{\pm} \quad \rightarrow \quad e^{\mp i\zeta} e^{\pm},$$
(3.2.8)

$$e^3 \quad \rightarrow \quad e^3 - d\zeta,$$
(3.2.9)

where ζ depends on the internal coordinates θ and φ, and the group element of $SU(2)$, associated to the rotation, but it does not depend on the 4D coordinates[4] x^μ. So the 1-form (3.2.2) has the following transformation property

$$A \rightarrow A + \frac{n}{2e} d\zeta.$$
(3.2.10)

We can now introduce the iso-helicity by saying that the iso-helicity of $e^{\pm}$ is ± 1. Further more, if we consider the background covariant derivative of ϕ and we remember this object must have the same iso-helicity of ϕ, we obtain that ϕ has iso-helicity $n/2$. Generally rotations act on tensors like an $SO(2)$ group, so we can group the components of tensors in $SO(2)$ irreducible pieces [7]: the iso-helicity of a field is nothing but its $SO(2)$ charge.

Generally if Φ_λ is a field with an integer or half-integer iso-helicity λ, we can perform an harmonic expansion [7]:

$$\Phi_\lambda(x,\theta,\phi) = \sum_{l\geq|\lambda|} \sum_{|m|\leq l} \Phi^l_m(x) \sqrt{\frac{2l+1}{4\pi}} \mathcal{D}^{(l)\lambda}_m(\theta,\varphi),$$
(3.2.11)

where, for a given l, $\mathcal{D}^{(l)\lambda}_m$ is a $(2l+1) \times (2l+1)$ unitary matrix. For example ϕ has an expansion like (3.2.11) with $\lambda = n/2$. The $\mathcal{D}^{(l)\lambda}_m$ were originally introduced in [50] and in the following we give our conventions. We define the harmonics $\mathcal{D}^{(l)\lambda}_m$ as proportional to the

---

[4]The transformation laws (3.2.8) and (3.2.9) can be extended to an x-dependent rotation [7].

matrix element

$$\langle l, \lambda | \, e^{i\varphi Q_3} e^{i(\pi - \theta) Q_2} e^{i\varphi Q_3} \, | l, m \rangle \,, \qquad (3.2.12)$$

where the Q_j, $j = 1, 2, 3$, are the generators of $SU(2)$:

$$[Q_j, Q_k] = i\epsilon_{jkl} Q_l, \qquad (3.2.13)$$

where ϵ_{jkl} is the totally antisymmetric Levi-Civita symbol with $\epsilon_{123} = 1$. Moreover $|l, m\rangle$ is the eigenvector of $\sum_j Q_j^2$ with eigenvalue $l(l+1)$ and the eigenvector of Q_3 with eigenvalue m. We introduce also $\mathcal{D}_{\lambda,m}^{(l)} \equiv \mathcal{D}_m^{(l)-\lambda}$. The explicit harmonic expansions for ϕ and the fluctuations h_{MN} and $\mathcal{V}_M$ of the metric and the gauge field are

$$\phi = \sum_{l \geq |n|/2} \sum_{|m| \leq l} \phi^l{}_m(x) \sqrt{\frac{2l+1}{4\pi}} \mathcal{D}_{-n/2,m}^{(l)}(\theta, \varphi), \qquad (3.2.14)$$

$$\mathcal{V}_\mu = \sum_{l \geq 0} \sum_{|m| \leq l} \mathcal{V}_\mu^l{}_m(x) \sqrt{\frac{2l+1}{4\pi}} \mathcal{D}_{0,m}^{(l)}(\theta, \varphi), \qquad (3.2.15)$$

$$h_{\mu+} = \sum_{l \geq 1} \sum_{|m| \leq l} h_{\mu+}^l{}_m(x) \sqrt{\frac{2l+1}{4\pi}} \mathcal{D}_{+,m}^{(l)}(\theta, \varphi), \qquad (3.2.16)$$

$$h_{\mu\nu} = \sum_{l \geq 0} \sum_{|m| \leq l} h_{\mu\nu}^l{}_m(x) \sqrt{\frac{2l+1}{4\pi}} \mathcal{D}_{0,m}^{(l)}(\theta, \varphi), \qquad (3.2.17)$$

$$\mathcal{V}_+ = \sum_{l \geq 1} \sum_{|m| \leq l} \mathcal{V}_+^l{}_m(x) \sqrt{\frac{2l+1}{4\pi}} \mathcal{D}_{+,m}^{(l)}(\theta, \varphi),$$

$$h_{++} = \sum_{l \geq 2} \sum_{|m| \leq l} h_{++}^l{}_m(x) \sqrt{\frac{2l+1}{4\pi}} \mathcal{D}_{2,m}^{(l)}(\theta, \varphi)$$

$$h_{+-} = \sum_{l \geq 0} \sum_{|m| \leq l} h_{+-}^l{}_m(x) \sqrt{\frac{2l+1}{4\pi}} \mathcal{D}_{0,m}^{(l)}(\theta, \varphi), \qquad (3.2.18)$$

where the subscripts $+$ and $-$ refer to the basis (3.2.6). For $l = 1$ our choice is

$$\mathcal{D}_{\hat\alpha,\hat\beta}(\theta,\varphi) = \begin{pmatrix} \frac{1}{2}(\cos\theta+1) & \frac{1}{2}(\cos\theta-1)e^{-2i\varphi} & -\frac{1}{\sqrt{2}}\sin\theta e^{-i\varphi} \\ \frac{1}{2}(\cos\theta-1)e^{2i\varphi} & \frac{1}{2}(\cos\theta+1) & -\frac{1}{\sqrt{2}}\sin\theta e^{i\varphi} \\ \frac{1}{\sqrt{2}}\sin\theta e^{i\varphi} & \frac{1}{\sqrt{2}}\sin\theta e^{-i\varphi} & \cos\theta \end{pmatrix},$$

$$(3.2.19)$$

where we have introduced $\mathcal{D}_{\lambda,m} \equiv \mathcal{D}^{(1)}_{\lambda,m}$. In (3.2.19) the first, second and third rows correspond to $\hat\alpha = +,-,3$, the first, second and third columns to $\hat\beta = +,-,3$. While our choice for $\mathcal{D}^{(2)}_{\lambda,m}$ is

$$\mathcal{D}^{(2)}_{\lambda,2}(\theta,\varphi) = \begin{pmatrix} \frac{1}{4}(1+\cos\theta)^2 \\ -\frac{1}{2}\sin\theta(1+\cos\theta)e^{i\varphi} \\ \sqrt{\frac{3}{8}}\sin^2\theta e^{2i\varphi} \\ -\frac{1}{2}\sin\theta(1-\cos\theta)e^{3i\varphi} \\ \frac{1}{4}(1-\cos\theta)^2 e^{4i\varphi} \end{pmatrix},$$

$$\mathcal{D}^{(2)}_{\lambda,1}(\theta,\varphi) = \begin{pmatrix} -\frac{1}{2}\sin\theta(1+\cos\theta)e^{-i\varphi} \\ \frac{1}{2}(1-\cos\theta-2\cos^2\theta) \\ \sqrt{\frac{3}{2}}\sin\theta\cos\theta e^{i\varphi} \\ \frac{1}{4}(4\cos^2\theta-2\cos\theta-2)e^{2i\varphi} \\ \frac{1}{2}\sin\theta(1-\cos\theta)e^{3i\varphi} \end{pmatrix},$$

$$\mathcal{D}^{(2)}_{\lambda,0}(\theta,\varphi) = \begin{pmatrix} \sqrt{\frac{3}{8}}\sin^2\theta e^{-2i\varphi} \\ \sqrt{\frac{3}{2}}\sin\theta\cos\theta e^{-i\varphi} \\ \frac{1}{2}(3\cos^2\theta-1) \\ -\sqrt{\frac{3}{2}}\sin\theta\cos\theta e^{i\varphi} \\ \sqrt{\frac{3}{8}}\sin^2\theta e^{2i\varphi} \end{pmatrix},$$

$$\mathcal{D}^{(2)}_{\lambda,-1}(\theta,\varphi) = \begin{pmatrix} -\frac{1}{2}\sin\theta(1-\cos\theta)e^{-3i\varphi} \\ \frac{1}{4}(4\cos^2\theta-2\cos\theta-2)e^{-2i\varphi} \\ -\sqrt{\frac{3}{2}}\sin\theta\cos\theta e^{-i\varphi} \\ \frac{1}{2}(1-\cos\theta-2\cos^2\theta) \\ \frac{1}{2}\sin\theta(1+\cos\theta)e^{i\varphi} \end{pmatrix},$$

$$\mathcal{D}^{(2)}_{\lambda,-2}(\theta,\varphi) = \begin{pmatrix} \frac{1}{4}\left(1-\cos\theta\right)^2 e^{-4i\varphi} \\ \frac{1}{2}\sin\theta(1-\cos\theta)e^{-3i\varphi} \\ \sqrt{\frac{3}{8}}\sin^2\theta e^{-2i\varphi} \\ \frac{1}{2}\sin\theta(1+\cos\theta)e^{-i\varphi} \\ \frac{1}{4}\left(1+\cos\theta\right)^2 \end{pmatrix},$$

where λ is a row index. We could continue and compute the harmonics for every value of l but we do not do that as we do not need their explicit expression for $l > 2$.

It is useful to compute the effect of the background covariant derivatives on the harmonics. We have

$$\nabla_\alpha \mathcal{D}^{(l)}_{\lambda,m} = e^n_\alpha \left(\partial_n - \lambda\omega_n\right)\mathcal{D}^{(l)}_{\lambda,m}, \tag{3.2.20}$$

where ∇_α is the background covariant derivative, e^n_α is the inverse of e^α_n, which can be calculated from (3.2.6), and ω_n represents the background spin connection: we have

$$\omega_\varphi = \frac{i}{a}(\cos\theta - 1), \quad \omega_\theta = 0. \tag{3.2.21}$$

It can be proved the following effects of

$$\nabla^2 = \nabla_\alpha\nabla^\alpha = \nabla_+\nabla_- + \nabla_-\nabla_+ \tag{3.2.22}$$

over the harmonics:

$$\nabla^2\mathcal{D}^{(l)}_{\lambda,m} = -\frac{1}{a^2}\left[l(l+1) - \lambda^2\right]\mathcal{D}^{(l)}_{\lambda,m}. \tag{3.2.23}$$

In the following we consider, just for simplicity, the case

$$n = 2. \tag{3.2.24}$$

In fact for this value of the monopole charge we can find a very simple solution of the fundamental 6D EOMs (3.1.8) which is invariant under a $U(1)$ subgroup of $SU(2) \times U(1)$; this solution is discussed in Section 3.3. Like in Section 2.2 our purpose is in fact to construct the 4D $SU(2) \times U(1)$-invariant effective theory, study the spontaneous symmetry breaking $SU(2) \times U(1) \to U(1)$ and the Higgs mechanism in the effective theory and then compare the results with the corresponding quantities predicted by the 6D theory; therefore, in order to do that, one has to find a 6D $U(1)$-invariant solution of the EOMs. We observe that for $n = 2$ the iso-helicity of ϕ is 1.

The low energy 4D spectrum coming from the background (3.2.1), (3.2.2) and (3.2.3) is given in Ref. [7] for the spin-1 and spin-2 sectors. The massless sector is the following: there are a graviton (helicities ± 2, $l = 0$), a $U(1)$ gauge field (helicities ± 1, $l = 0$) coming from $\mathcal{V}_\mu$ and a Yang-Mills $SU(2)$ triplet (helicities ± 1, $l = 1$) coming from $h_{\mu\alpha}$ and $\mathcal{V}_\mu$, where $\mathcal{V}_M$ and h_{MN} are the fluctuations of the gauge field and the metric around solution (3.2.1), (3.2.2) and (3.2.3). Regarding the scalar spectrum all the scalars from G_{MN} and A_M have very large masses, of the order $1/a$, and we can get only an $SU(2)$-triplet with mass squared μ^2 from ϕ in the low energy spectrum if we choose m^2 such that

$$|\mu^2| \ll \frac{1}{a^2}, \qquad (3.2.25)$$

where

$$\mu^2 \equiv -\frac{1}{a^2}\eta \equiv m^2 + \frac{1}{a^2}. \qquad (3.2.26)$$

In fact $-1/a^2$ is the eigenvalue of the Laplacian operator acting on the harmonic with $l = 1$ and $\lambda = 1$, as one can check using the related formula of [7]. The parameter μ^2 is in fact the squared mass of the triplet from ϕ, and it can be in principle either positive or negative. If (3.2.25) holds all the remaining scalars have masses at least of the order $1/a$ and they do not appear in the low energy theory. So we assume that (3.2.25) holds. Finally in order to find the low energy fermionic spectrum we have to calculate the associated iso-helicities by using the explicit expression for the background covariant derivative of $\psi_\pm$ along the internal space:

$$\nabla_m \psi_\pm = \left(\partial_m \pm \omega_m \frac{1}{2}\gamma^5 + ie_\pm A_m \right) \psi_\pm, \qquad (3.2.27)$$

where $\omega_\theta = 0$, $\omega_\varphi = \frac{i}{a}(\cos\theta - 1)$, $e_+ = e/2$ and $e_- = 3e/2$. We get

$$\lambda_{+L} = 0, \quad \lambda_{+R} = 1, \quad \lambda_{-L} = 2, \quad \lambda_{-R} = 1 \qquad (3.2.28)$$

and the corresponding expansions are given by (3.2.11). So the equations (3.1.12) tell us that there are 4 zero-modes: the $l = 0$, $m = 0$ mode in ψ_{+L} and the $l = 1$, $m = +1, -1, 0$ in ψ_{-R}. So we have a massless $SU(2)$ singlet from ψ_{+L} and a massless $SU(2)$ triplet from ψ_{-R}.

3.2.2 The 4D $SU(2) \times U(1)$ Effective Lagrangian and the Higgs Mechanism

Now we want to study the 4D effective theory: which is the 4D theory obtained from the background (3.2.1), (3.2.2) and (3.2.3) retaining only the low energy spectrum we discussed at the end of Subsection 3.2.1, that is the particles with masses much smaller than $1/a$, and integrating out all the heavy modes, namely those with mass at least of the order $1/a$. This is an $SU(2) \times U(1)$-invariant theory, which includes a charged scalar, that we call χ, in the 3-dimensional representation of $SU(2)$, and, if we want, two Weyl spinors in the $1_{1/2}$ and $3_{3/2}$ of $SU(2) \times U(1)$. The background (3.2.1), (3.2.2) and (3.2.3) is the analogous of what we called Φ_{eff} in Section 2.2. In this section we give only some relevant terms[5] appearing in the lagrangian of this theory. In particular we calculate the scalar potential, we study the Higgs mechanism, which is active only for $\mu^2 < 0$, and we give in this case the masses of the spin-1, spin-0 and spin-1/2 particles.

Like in the general scalar theory of Section 2.2, in the following we perform all the calculations at the order η. If we use the information regarding the low energy spectrum which we discussed at the end of Subsection 3.2.1, we can construct some relevant terms of the 4D effective theory through the following ansatz[6]

$$
\begin{aligned}
E^a(x) &= E^a_\mu(x)dx^\mu, \\
E^\alpha(x,y) &= e^\alpha(y) - \frac{\kappa}{a\sqrt{4\pi}}W^{\hat\alpha}_\mu(x)dx^\mu \mathcal{D}^\alpha_{\hat\alpha}(y), \\
A(x,y) &= -\frac{n}{2ea}e^3(y) \\
&\quad + \frac{1}{a\sqrt{4\pi}}V_\mu(x)dx^\mu - \frac{n\kappa}{2ea^2\sqrt{4\pi}}U^{\hat\alpha}_\mu(x)dx^\mu \mathcal{D}^3_{\hat\alpha}(y), \\
\phi(x,y) &= \frac{1}{a}\sqrt{\frac{3}{4\pi}}\chi^{\hat\alpha}(x)\mathcal{D}_{-,\hat\alpha}(y), \\
\psi_{+R} &= \psi_{-L} = 0, \\
\psi_{-R} &= \frac{1}{a}\sqrt{\frac{3}{4\pi}}\psi^{\hat\alpha}_R(x)\mathcal{D}_{-,\hat\alpha}(y), \\
\psi_{+L} &= \frac{1}{a\sqrt{4\pi}}\psi_L(x), \quad\quad\quad\quad\quad\quad\quad (3.2.29)
\end{aligned}
$$

---

[5]Here "relevant terms" has the same meaning as in the Subsection 2.2.1.
[6]The ansatz (3.2.29) is a generalization of the zero-mode ansatz of [7], which does not include scalar fields.

where E^A, $A = 0, 1, 2, 3, +, -$, are the 6D (x, y)-dependent orthonormal 1-form basis, E^a_μ is the 4D x-dependent vielbein, V_μ is the 4D $U(1)$ gauge field coming from $\mathcal{V}_\mu$, a linear combination[7] of W_μ and U_μ is the Yang-Mills $SU(2)$ triplet [7] coming from $h_{\mu\alpha}$ and $\mathcal{V}_\mu$; finally ψ_L and ψ_R are the $SU(2)$ fermion singlet and fermion triplet, respectively. Actually the ansatz (3.2.29) is the background (3.2.1), (3.2.2) and (3.2.3) plus some fluctuations, which include all the light KK states.

Now we want to write some relevant terms of the effective lagrangian for χ by using the light-mode ansatz (3.2.29) and by taking into account the heavy mode contribution. Concerning the scalar potential in the 4D effective theory, we already know that the bilinear part is simply $\mu^2 \chi^\dagger \chi$. Whereas the quartic terms are non-trivial and they have two different contributions: the quartic term in the 6D potential V in (3.1.4) computed with $\phi(x, y)$ in Eq. (3.2.29) and the heavy scalar modes ($h_{\alpha\beta}$ and $\mathcal{V}_\alpha$) contribution through diagrams like Fig. 2.1, evaluated at transferred momentum equal to zero. The first contribution is

$$\xi \int a^2 \sin\theta d\theta d\varphi \, |\phi(x, \theta, \phi)|^4 \, . \qquad (3.2.30)$$

This object is equal to

$$\frac{\xi}{a^2} \left(\chi_{m_1} \right)^* \chi_{m_2} \left(\chi_{m_3} \right)^* \chi_{m_4} J_{m_1 m_2 m_3 m_4}, \qquad (3.2.31)$$

where $J_{m_1 m_2 m_3 m_4}$ is an invariant tensor in the $3 \times 3 \times 3 \times 3$ representation of $SU(2)$. We obtain

$$J_{\hat{a}_1 \hat{a}_2 \hat{a}_3 \hat{a}_4} = j_1 \delta_{\hat{a}_1 \hat{a}_2} \delta_{\hat{a}_3 \hat{a}_4} + j_2 g_{\hat{a}_1 \hat{a}_3} g_{\hat{a}_2 \hat{a}_4} + j_3 \delta_{\hat{a}_1 \hat{a}_4} \delta_{\hat{a}_2 \hat{a}_3}, \qquad (3.2.32)$$

where j_1, j_2 and j_3 are some constants. By explicit calculations we get

$$j_1 + j_3 = \frac{9}{20\pi}, \quad j_2 = -\frac{3}{20\pi}. \qquad (3.2.33)$$

The final expression for the scalar potential $\mathcal{U}$ in the 4D effective theory, including the bilinear and the quartic interactions and the light

---

[7]The orthogonal linear combination has a large mass; we show this in Appendix B.1.3.

and heavy mode contributions, is

$$\mathcal{U}(\chi) = \mu^2 \chi^\dagger \chi + (\lambda_H + c_1 \lambda_G)\left(\chi^\dagger \chi\right)^2 - \frac{\lambda_H + c_2 \lambda_G}{3}\left|\chi^{\hat{\alpha}} g_{\hat{\alpha}\hat{\beta}} \chi^{\hat{\beta}}\right|^2 + ...,$$

$$(3.2.34)$$

where c_1 and c_2 are dimensionless parameters,

$$\lambda_H \equiv \frac{9}{20\pi a^2}\xi, \qquad \lambda_G \equiv \frac{9\kappa^2}{80\pi a^4} \qquad (3.2.35)$$

and the dots represent higher order non relevant terms, for example terms with a product of 6 χ or 8 χ. These terms do not contribute to the VEV of χ as we want this VEV to be of the order[8] $\eta^{1/2}$. In (3.2.34) the contribution of the heavy scalars, namely $h_{\alpha\beta}$ and V_α, is represented by $c_1\lambda_G$ and $c_2\lambda_G$, the analogous of a_{lmpq} in the Eq. (2.2.10). Moreover we give also the expression for the gauge covariant derivative of χ:

$$D_\mu \chi^{\hat{\alpha}} = \partial_\mu \chi^{\hat{\alpha}} + ig_1 V_\mu \chi^{\hat{\alpha}} + g_2 \mathcal{A}_\mu^{\hat{\beta}} \epsilon_{\hat{\beta}\hat{\gamma}}{}^{\hat{\alpha}} \chi^{\hat{\gamma}},$$

$$(3.2.36)$$

where $\mathcal{A}_\mu$ is defined in Appendix B.1.3 and it represents the $SU(2)$ Yang-Mills field, $\epsilon_{\hat{\gamma}\hat{\beta}\hat{\alpha}}$ is a totally antisymmetric symbol with $\epsilon_{+-3} = i$, and

$$g_1 = \frac{e}{\sqrt{4\pi a}}, \qquad g_2 = \sqrt{\frac{3}{16\pi}}\frac{\kappa}{a^2}, \qquad (3.2.37)$$

are the 4D $U(1)$ and $SU(2)$ gauge couplings. Therefore the complete lagrangian for χ is

$$\mathcal{L}_{\chi eff} = -\left(D_\mu \chi\right)^\dagger D^\mu \chi - \mathcal{U}(\chi). \qquad (3.2.38)$$

Let us look for the points of minimum of the order $\eta^{1/2}$ of the potential $\mathcal{U}$ in (3.2.34). We have a minimum, in the case $\mu^2 < 0$, for

$$\chi_1 = \chi_2 = 0, \quad \chi_3 = v \equiv \sqrt{\frac{-3\mu^2}{4\left[\lambda_H + \frac{1}{2}\left(3c_1 - c_2\right)\lambda_G\right]}}, \qquad (3.2.39)$$

which corresponds to the global minimum

$$\mathcal{U}_0 = 0 \qquad (3.2.40)$$

---

[8]The order $\eta^{1/2}$ corresponds to the order μ because of Eq. (3.2.26).

at the order η. This fact states that, at leading order, the 4D flatness condition in the background is compatible with the procedure of the 4D effective theory. In fact $\mathcal{U}_0$ can be interpreted as a 4D cosmological constant and the flatness implies $\mathcal{U}_0 = 0$. Instead for $\mu^2 > 0$ we do not have any order parameter because the global minimum $\mathcal{U}_0 = 0$ corresponds to $\chi = 0$.

If we take, for $\mu^2 < 0$, the vacuum (3.2.39), $SU(2) \times U(1)$ breaks to $U(1)_3$, where $U(1)_3$ is the $U(1)$-subgroup of $SU(2)$ generated by its third generator. The gauge field of $U(1)$ and $SU(2)$ are respectively V_μ and $\mathcal{A}_\mu$; before Higgs mechanism these gauge fields are of course massless as one can see by looking at their bilinear lagrangian given in Appendix B.1.3. From (3.2.38) and (3.2.36) we can calculate the masses of these vector fields in the 4D effective theory after the Higgs mechanism. We get a massless vector field $\mathcal{A}_\mu^3$, which corresponds to the unbroken $U(1)_3$ gauge symmetry. Instead V_μ and $\mathcal{A}_\mu^\pm$ acquire respectively the following squared masses

$$M_V^2 = \frac{3e^2}{8\pi a^2} \frac{-\mu^2}{\lambda_H + \frac{1}{2}(3c_1 - c_2)\lambda_G}, \qquad (3.2.41)$$

$$M_{V\pm}^2 = \frac{9e^2}{16\pi a^2} \frac{-\mu^2}{\lambda_H + \frac{1}{2}(3c_1 - c_2)\lambda_G}, \qquad (3.2.42)$$

where the subscript V indicates that we are dealing with vector particles. Moreover, in the spin-0 sector, we have two physical scalar fields: a real scalar and a complex one, which is charged under the residual $U(1)_3$ symmetry. Their squared masses are respectively

$$M_S^2 = -2\mu^2, \qquad (3.2.43)$$

$$M_{S\pm}^2 = -\mu^2 \frac{\lambda_H + c_2\lambda_G}{\lambda_H + \frac{1}{2}(3c_1 - c_2)\lambda_G}. \qquad (3.2.44)$$

Finally we can determine the fermionic spectrum by examining the fermionic lagrangian in the effective theory:

$$\mathcal{L}_{Feff} = \overline{\psi_L}\gamma^\mu D_\mu \psi_L + \overline{\psi_R}\gamma^\mu D_\mu \psi_R + g_4\overline{\psi_L}\chi^\dagger\psi_R + g_4\overline{\psi_R}\chi\psi_L, \qquad (3.2.45)$$

where

$$g_4 = \frac{g_Y}{a\sqrt{4\pi}}. \qquad (3.2.46)$$

The result is a neutral Dirac fermion, with squared mass

$$M_F^2 = \frac{3g_Y^2}{16\pi a^2} \frac{-\mu^2}{\lambda_H + \frac{1}{2}(3c_1 - c_2)\lambda_G},$$ (3.2.47)

and a pair of massless right-handed Weyl fermions. We observe that the mass spectrum that we gave here is parametrized by the c_i. Of course these constants are not free parameters but they can be in principle computed by evaluating explicitly the heavy modes contribution. In the rest of the present chapter we do not compute the c_i but we prove that the 4D effective theory without heavy mode contribution, that is $c_i = 0$, is not correct because it predicts a wrong VEV of the light KK scalars and a wrong mass spectrum.

3.3   6D Electroweak Symmetry Breaking

Now we perform a 6D (or geometrical) analysis of spontaneous symmetry breaking: this method corresponds to the contents of Section 2.2 for scalar theories. Of course we perform all the calculations at the order η, as in the effective theory method. So our first purpose is finding a solution of the 6D EOM which breaks the $SU(2) \times U(1)$ symmetry at the 6D level and which is a small perturbation of the order $\eta^{1/2}$ of the sphere solution (3.2.1), (3.2.2) and (3.2.3).

In order to find such a solution we consider an expansion of all background tensors in powers of $\eta^{1/2}$. For the ansatz (3.1.5), (3.1.6) and (3.1.7) our tensors are g_{mn}, A_m and ϕ; the expansion of the latter is

$$\phi = \sum_{k=1}^{\infty} \phi_k \eta^{k/2}.$$ (3.3.1)

We have omitted the $k = 0$ term because we want that ϕ goes to zero as η goes to zero. Now we are interested in the EOM for ϕ, namely the first equation of (3.1.8). Since that equation involves also the Laplacian ∇^2 acting on charged scalar, we expand also this operator in powers of $\eta^{1/2}$:

$$\nabla^2 = \nabla_0^2 + \sum_{k=1}^{\infty} L_k,$$ (3.3.2)

where ∇_0^2 is the Laplacian corresponding to the $SU(2) \times U(1)$-invariant solution (3.2.1) and (3.2.2) and L_k is an operator proportional to $\eta^{k/2}$.

By putting (3.3.1) and (3.3.2) in the first equation of (3.1.8) we get one equation for every power of $\eta^{1/2}$. The first one is

$$\eta^{1/2}\left(\nabla_0^2 + \frac{1}{a^2}\right)\phi_1 = 0, \qquad (3.3.3)$$

which implies that ϕ_1 must be proportional to the harmonic with $l = 1$ and $\lambda = 1$. Further we impose $m = 0$, otherwise we do not have an $U(1)_3$-invariant background. So we have

$$\phi_1 \propto \mathcal{D}_0^{(1)1} \equiv D. \qquad (3.3.4)$$

Moreover the equation proportional to η is

$$\eta\left(\nabla_0^2 + \frac{1}{a^2}\right)\phi_2 + \eta^{1/2}L_1\phi_1 = 0. \qquad (3.3.5)$$

On the other hand the operator L_1 must vanish because from (3.1.8) follows

$$\frac{1}{k^2}R_{mn} - g_{mn}\left(\lambda + m^2|\phi|^2 + \xi|\phi|^4\right) - \frac{1}{2}(\nabla_m\phi)^*\nabla_n\phi - \frac{1}{2}(\nabla_n\phi)^*\nabla_m\phi = 0, \qquad (3.3.6)$$

which, up to $O(\eta)$, reduces to

$$\frac{1}{k^2}R_{mn} - g_{mn}\lambda = 0. \qquad (3.3.7)$$

Since the only solution of (3.3.7) is the round S^2, there is no $\eta^{1/2}$ terms in g_{mn}. By putting this result in the last equation of (3.1.8) we get that also the gauge field A_m cannot have $\eta^{1/2}$ terms. So we have $\mathcal{O}_1 = 0$ and (3.3.5) becomes

$$\left(\nabla_0^2 + \frac{1}{a^2}\right)\phi_2 = 0, \qquad (3.3.8)$$

which, taking into account also the $U(1)_3$-invariance, implies

$$\phi_2 \propto D. \qquad (3.3.9)$$

For simplicity we take $\phi_2 = 0$ because in any case ϕ_2 must be proportional to the same harmonic of ϕ_1. So $\phi = \eta^{1/2}\phi_1$ up to $O(\eta^{3/2})$. The equation proportional to $\eta^{3/2}$ is then

$$\eta^{3/2}\left(\nabla_0^2 + \frac{1}{a^2}\right)\phi_3 + \eta^{1/2}L_2\phi_1 + \eta^{3/2}\frac{1}{a^2}\phi_1 = 2\xi\eta^{1/2}|\eta||\phi_1|^2\phi_1.$$
(3.3.10)

By projecting this equation over the harmonic D, the first term disappears and we get

$$\frac{1}{a^2}\eta^{3/2}\int D^*\phi_1 + \eta^{1/2}\int D^*L_2\phi_1 = \eta^{1/2}|\eta|2\xi\int D^*|\phi_1|^2\phi_1. \quad (3.3.11)$$

The most simple solution of this kind, up to higher order terms in η, that we find is similar to the background which appears in Ref. [51] [9]:

$$
\begin{aligned}
ds^2 &= \eta_{\mu\nu}dx^\mu dx^\nu + a^2\left[(1 + |\eta|\beta\sin^2\theta)d\theta^2 + \sin^2\theta d\varphi^2\right], \\
A &= -\frac{1}{e}e^3, \\
\phi &= \eta^{1/2}\alpha\exp\left(i\varphi\right)\sin\theta,
\end{aligned}
\quad (3.3.12)
$$

where $\beta \equiv \kappa^2|\alpha|^2$. As required, for $\eta = 0$ this background reduces to the background of Subsection 3.2.1. The absolute value of α can be computed by using Eq. (3.3.11), which, through the redefinition $\eta^{1/2}\phi_1 \to \phi$ reads

$$\frac{1}{a^2}\eta\int D^*\phi + \int D^*L_2\phi = 2\xi\int D^*|\phi|^2\phi. \quad (3.3.13)$$

The metric appearing in solution (3.3.12) is the metric of an ellipsoid. From the geometrical point of view we have deformed our internal space as shown in Fig. 3.1.

For $\mu^2 < 0$ the equation (3.3.13) has a solution for

$$\lambda_H > \lambda_G, \quad (3.3.14)$$

where λ_H and λ_G are defined by (3.2.35), while, for $\mu^2 > 0$, we have a solution for

$$\lambda_H < \lambda_G. \quad (3.3.15)$$

---

[9]This solution was discussed in Ref. [51], but incorrectly.

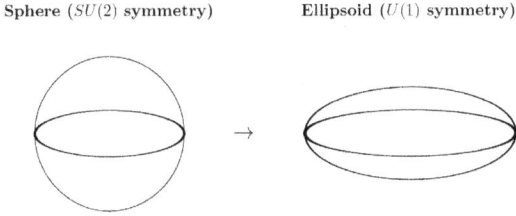

Sphere ($SU(2)$ symmetry)   Ellipsoid ($U(1)$ symmetry)

Figure 3.1: We show the deformation of the internal space in the 6D approach to the electroweak symmetry breaking. The elecroweak gauge symmetry is broken to $U(1)$ through the ellipsoid background.

Whether $\mu^2 > 0$ or $\mu^2 < 0$, the solution of (3.3.13) is

$$|\alpha|^2 = \frac{5}{|8\xi a^2 - 2\kappa^2|} = \frac{9}{32\pi a^4} \frac{1}{|\lambda_H - \lambda_G|}. \qquad (3.3.16)$$

Note that here we have symmetry breaking for both signs of μ^2. This is not so interesting because the solution with $\mu^2 > 0$ is unstable, as it is discussed in Subsection 3.3.2. We want to stress that the value of $|\alpha|^2$ predicted by the 4D effective theory is not equal to (3.3.16) if we neglect the heavy mode contribution to the effective theory, namely for $c_i = 0$: indeed in this case the effective theory predicts a value of $|\alpha|^2$ equal to

$$|\alpha|^2_{eff} = \frac{9}{32\pi a^4} \frac{1}{\lambda_H}, \qquad (3.3.17)$$

which is equal to (3.3.16) only for $\lambda_G = 0$. However, from (3.2.35) it's clear that λ_G cannot be taken equal to zero. Therefore we have already proved that the heavy modes contribution is needed at least for the light mode VEV. We shall prove that this is the case also for the mass spectrum.

As required the background (3.3.12) has the symmetry

$$U(1)_3 \subset SU(2). \qquad (3.3.18)$$

So the 4D effective low energy theory, which follows from this background, is $U(1)_3$-invariant and comparing these results with the effective theory predictions makes sense.

We note that the symmetry breaking (3.3.18) is associated, in the 6D theory, to a geometrical deformation of the internal space. Further we observe that (3.3.12) tell us the heavy modes VEVs are higher order corrections with respect to the light modes VEVs like in the scalar theories of section 2.2.

Now we calculate the low energy vector, scalar and fermion spectrum by analyzing the 4D bilinear lagrangian for the fluctuations around the solution (3.3.12).

3.3.1 Spin-1 Spectrum

The spin-1 spectrum can be calculated in a way similar to the light mode ansatz (3.2.29). However, it must be noted that the sectors with different l no longer decouple for $\eta \neq 0$, but the mixing terms are of the order η and they give negligible corrections of the order η^2 to the vector boson masses. These facts are evident from the general formula of [52]. So we can neglect the modes with $l > 1$ in the calculation of spin-1 spectrum. Therefore we can compute the vector boson masses by putting the following ansatz in the action and integrating over the extra dimensions:

$$
\begin{aligned}
E^a(x) &= E^a_\mu(x)dx^\mu, \\
E^\alpha(x,y) &= e^\alpha(y,\eta) - \frac{\kappa}{a\sqrt{4\pi}} W^{\hat{\alpha}}_\mu(x)dx^\mu \mathcal{D}^\alpha_{\hat{\alpha}}(y), \\
A(x,y) &= -\frac{1}{ea}e^3(y) \\
&\quad + \frac{1}{a\sqrt{4\pi}} V_\mu(x)dx^\mu - \frac{\kappa}{ea^2\sqrt{4\pi}} U^{\hat{\alpha}}_\mu(x)dx^\mu \mathcal{D}^3_{\hat{\alpha}}(y), \\
\phi(x,y) &= \eta^{1/2}\alpha \exp(i\varphi)\sin\theta, \quad\quad (3.3.19)
\end{aligned}
$$

where $e^\alpha(y,\eta)$ is the orthonormal basis for the 2-dimensional metric in (3.3.12): :

$$
e^\pm(y,\eta) = \pm\frac{i}{\sqrt{2}}e^{\pm i\varphi}\left[\left(1 + |\eta|\frac{\beta}{2}\sin^2\theta\right)d\theta \pm i\sin\theta d\varphi\right]. \quad (3.3.20)
$$

In (3.3.19) we consider the spin-1 fluctuations but we do not consider the spin-0 fluctuations, because they are not necessary for the calculation of vector boson masses. It's important to note that in (3.3.19) the VEV of E^α is $e^\alpha(y,\eta)$, it's not $e^\alpha(y)$ as in (3.2.29).

From (3.3.19) it follows that some of the previous ($\eta = 0$) massless states acquire masses for $\eta \neq 0$. Up to $O(\eta^{3/2})$, the $U(1)$ gauge boson

$(l = 0)$ has the mass squared

$$M_V^2 = \eta \frac{20}{3} \frac{e^2}{8\xi a^2 - 2\kappa^2} = \frac{3e^2}{8\pi a^2} \frac{-\mu^2}{\lambda_H - \lambda_G}, \qquad (3.3.21)$$

while the Yang-Mills triplet $\mathcal{A}$ $(l = 1)$ is separated in a massless gauge boson, which is associated to $U(1)_3$ gauge invariance, and a couple of massive vector fields with the same mass squared

$$M_{V\pm}^2 = \eta \frac{10e^2}{8\xi a^2 - 2\kappa^2} = \frac{9e^2}{16\pi a^2} \frac{-\mu^2}{\lambda_H - \lambda_G}. \qquad (3.3.22)$$

By comparing (3.3.21) and (3.3.22) with (3.2.41) and (3.2.42), we get that the heavy mode contribution is needed in the effective theory. However, we observe that the ratio $M_V^2/M_{V\pm}^2$ is correctly predicted by the 4D effective theory for every c_i.

Since the computation of vector bosons masses is complicated we present it explicitly. In order to prove (3.3.21) and (3.3.22) it's useful to split the action in four terms:

$$S_B = S_R + S_F + S_\lambda + S_\phi, \qquad (3.3.23)$$

where

$$S_R = \int d^6 X \sqrt{-G} \frac{1}{\kappa^2} R, \qquad (3.3.24)$$

$$S_F = -\frac{1}{4} \int d^6 X \sqrt{-G} F^2, \qquad (3.3.25)$$

$$S_\lambda = \int d^6 X \sqrt{-G} \, (-\lambda), \qquad (3.3.26)$$

$$S_\phi = \int d^6 X \sqrt{-G} \left[-(\nabla_M \phi)^* \nabla^M \phi - V(\phi) \right]. \qquad (3.3.27)$$

In Appendix B.1 we prove that the contributions coming from S_R and S_F vanish, so only S_ϕ contributes to the spin-1 masses up to $O(\eta^{3/2})$. The same low energy spin-1 masses in (3.3.21) and (3.3.22) can be obtained also by using the general formula of [52], which contains all the bilinear terms in the light cone gauge. The light cone gauge advantage is that the sectors with different spin decouple. However, the derivation that we presented here shows that the unique contribution (at the leading order) to the spin-1 masses comes from S_ϕ, like in the effective theory approach. This explains why the ratio $M_V^2/M_{V\pm}^2$ is correctly predicted by the 4D effective theory for every values of c_i.

3.3.2 Spin-0 Spectrum

We choose the light cone gauge [52, 53, 54] in order to evaluate the spin-0 spectrum. In this gauge we have just two independent values for the indexes $\mu, \nu, \ldots$ which label the 4D coordinates. The bilinears for the fluctuations over the solution (3.3.12) can be simply computed with the general formula of [52]. For our model the helicity-0 $\mathcal{L}_0$ part is given by

$$\mathcal{L}_0 = \mathcal{L}_0(\phi, \phi) + \mathcal{L}_0(h, h) + \mathcal{L}_0(\mathcal{V}, \mathcal{V}) + \mathcal{L}_0(\phi, h) + \mathcal{L}_0(\phi, \mathcal{V}) + \mathcal{L}_0(h, \mathcal{V}),$$
(3.3.28)

where

$$\mathcal{L}_0(\phi, \phi) = \phi^* \partial^2 \phi + \phi^* \nabla^2 \phi - \left[m^2 + (4\xi + e^2)|\Phi|^2 \right.$$
$$+ \kappa^2 \left(\nabla_m \Phi \right)^* \nabla^m \Phi \right] |\phi|^2 - \frac{1}{2} \left\{ \left[(2\xi - e^2) \left(\Phi^* \right)^2 \right. \right.$$
$$\left. + \kappa^2 \left(\nabla_m \Phi \nabla^m \Phi \right)^* \right] \phi^2 + c.c. \right\},$$
(3.3.29)

$$\mathcal{L}_0(h, h) = \frac{1}{4\kappa^2} \left\{ h_{mn} \partial^2 h^{mn} + h_{mn} \nabla^2 h^{mn} + 2R_{mn}{}^{kl} h_l^m h_k^n \right.$$
$$+ \kappa^2 h_{ks} h_{mn} F^{km} F^{sn} - 2\kappa^2 h_m^l h_{ln} \left[\frac{1}{2} F^m_k F^{nk} + (\nabla^m \Phi)^* \nabla^n \Phi \right]$$
$$\left. + \frac{1}{2} h_i^j \partial^2 h_j^i + \frac{1}{2} h_i^j \nabla^2 h_j^i \right\},$$
(3.3.30)

$$\mathcal{L}_0(\mathcal{V}, \mathcal{V}) = \frac{1}{2} \left\{ \mathcal{V}_m \partial^2 \mathcal{V}^m + \mathcal{V}_m \nabla^2 \mathcal{V}^m - R_{mn} \mathcal{V}^m \mathcal{V}^n \right.$$
$$\left. - 2e^2 |\Phi|^2 \mathcal{V}^m \mathcal{V}_m - \kappa^2 \left(F_{ml} \mathcal{V}^l \right)^2 \right\},$$
(3.3.31)

$$\mathcal{L}_0(\phi, h) = \nabla_l h^{lm} \phi^* \nabla_m \Phi + h^{mn} \left(\nabla_m \phi \right)^* \nabla_n \Phi + c.c.,$$
(3.3.32)

$$\mathcal{L}_0(\phi, \mathcal{V}) = 2ie \mathcal{V}^m \phi^* \nabla_m \Phi - \kappa^2 F^{lm} \mathcal{V}_m \phi^* \nabla_l \Phi + c.c.,$$
(3.3.33)

$$\mathcal{L}_0(h, \mathcal{V}) = \mathcal{V}^n \left(\nabla_m h_{ln} F^{lm} - h_l^m \nabla_m F^l{}_n \right),$$
(3.3.34)

where Φ and ϕ are the background and the fluctuation of the 6D scalar. In this vanishing-helicity sector, it turns out that we have not only mixing terms of the order η but also mixing terms of the order $\eta^{1/2}$, coming from $\mathcal{L}_0(\phi, h)$ and $\mathcal{L}_0(\phi, \mathcal{V})$. So now we can't neglect the mixing between the sectors with different values of l, as we did in the helicity ± 1 sector. If we integrate these bilinear terms over the extra-dimensions we get an infinite dimensional squared mass matrix. However, we are interested only in the light masses, therefore we can use the perturbation theory of quantum mechanics in order to extract the correction of the order η to the masses of the 6 real scalars which are massless for $\eta = 0$. We already used this method for the computation of the mass spectrum in the scalar theories of Section 2.2. We explain now how to use it in this framework.

Formally we can write the bilinears $\mathcal{L}_0$ of the scalar fields in this way

$$\mathcal{L}_0 = \frac{1}{2} S^\dagger \partial^2 S - \frac{1}{2} S^\dagger \mathcal{O} S, \qquad (3.3.35)$$

where S is an array which includes all the scalar fluctuations; we choose

$$S = \begin{pmatrix} \phi \\ \phi^* \\ h_{++} \\ h_{--} \\ h_{+-} \\ \mathcal{V}_+ \\ \mathcal{V}_- \end{pmatrix}. \qquad (3.3.36)$$

We have just to solve a 2-dimensional eigenvalue problem for the squared mass operator[10] $\mathcal{O}$:

$$\mathcal{O} S = M^2 S. \qquad (3.3.37)$$

In particular we want to find the 6 values of M^2 which go to zero as η goes to zero. Since we are working at the order η we decompose $\mathcal{O}$ as follows

$$\mathcal{O} = \mathcal{O}_0 + \mathcal{O}_1 + \mathcal{O}_2, \qquad (3.3.38)$$

where $\mathcal{O}_0$ does not depend on η, $\mathcal{O}_1$ is proportional to $\eta^{1/2}$ and $\mathcal{O}_2$ is proportional to η. From the perturbation theory of quantum mechanics in the degenerate case we know that the 6 values of M^2 we

---

[10]The matrix elements of $\mathcal{O}$ can be computed by comparing (3.3.35) with the explicit expression of $\mathcal{L}_0$.

are interested in are the eigenvalues of the following 6×6 matrix[11]:

$$M^2_{ij} = -\sum_{\tilde{i}} \frac{< i|\mathcal{O}_1|\tilde{i} >< \tilde{i}|\mathcal{O}_1|j >}{M^2_{\tilde{i}}} + < i|\mathcal{O}_2|j >, \qquad (3.3.39)$$

where $|i >$, $i = 1, ...6$ represent the 6 orthonormal eigenfunctions of $\mathcal{O}_0$ with vanishing eigenvalue and they have the form

$$|i >= \begin{pmatrix} \phi \\ \phi^* \\ 0 \\ . \\ . \\ . \\ 0 \end{pmatrix}. \qquad (3.3.40)$$

Moreover $|\tilde{i} >$ are all the remaining orthonormal eigenfunctions of $\mathcal{O}_0$ and $M^2_{\tilde{i}}$ the corresponding eigenvalues. We note that the matrix elements $< i|\mathcal{O}_1|\tilde{i} >$ are non vanishing for

$$|\tilde{i} >= \begin{pmatrix} 0 \\ 0 \\ h_{++} \\ h_{--} \\ h_{+-} \\ \mathcal{V}_+ \\ \mathcal{V}_- \end{pmatrix}. \qquad (3.3.41)$$

Further the operator $\mathcal{O}_1$ modifies the integration measure just by a factor proportional to the harmonics $\mathcal{D}^{(1)}$, therefore we need just a finite subset of $|\tilde{i} >$ for the evaluation of M^2_{ij}, namely those constructed through the harmonics with $l = 0, 1, 2$, which are given in Appendix A. An explicit form for $|i >$ and $|\tilde{i} >$, and the preliminary computations of the 6 eigenvalues we are interested in, are given in Appendix B.2.

We give here just the final result: we have two unphysical scalar fields (a real and a complex one) which form the helicity-0 component of the massive vector fields; they have in fact the same squared masses

---

[11]Like in Section 2.2 we use the Dirac notation; for two states $|S_1 >$ and $|S_2 >$ and for an operator A, $< S_1|A|S_2 >$ represents $\int S^\dagger_1 A S_2$, where the integral is performed with the round S^2 metric.

given in (3.3.21) and (3.3.22), as it's required by Lorentz invariance, which is not manifest in the light cone gauge. Then we have a physical real scalar and a physical complex scalar, charged under the residual U(1) symmetry, with squared masses given respectively by (for $\mu^2 < 0$)

$$
\begin{aligned}
M_S^2 &= -2\mu^2, \\
M_{S\pm}^2 &= -\mu^2 \frac{\lambda_H + \lambda_G}{\lambda_H - \lambda_G}.
\end{aligned}
\tag{3.3.42}
$$

For $\mu^2 > 0$, we get a negative value for M_S^2, therefore the corresponding solution is unstable. Note that the squared mass M_S^2 has exactly the same expression as in the 4D effective theory, for every c_i. But for $c_i = 0$, which corresponds to neglecting the heavy mode contribution, the effective theory prediction for $M_{S\pm}^2$ in (3.2.44) is not equal to the correct value (3.3.42). We note that this is a physical inequivalence because the ratio $M_S^2/M_{S\pm}^2$, which is in principle a measurable quantity, is not correctly predicted by the 4D effective theory without the heavy mode contribution. More precisely the effective theory prediction for $M_S^2/M_{S\pm}^2$, in the case $c_i = 0$, is always greater than the correct value.

3.3.3 Spin-1/2 Spectrum

The spin-1/2 spectrum can be calculated by linearizing the EOM (3.1.12): for $n = 2$ we get

$$
\begin{aligned}
&\left(\partial^2 + 2\nabla_+\nabla_- - g_Y^2 |\Phi|^2\right)\psi_{+L} = 0, \\
&\left(\partial^2 + 2\nabla_-\nabla_+ - g_Y^2 |\Phi|^2\right)\psi_{-L} = 0, \\
&\left(\partial^2 + 2\nabla_-\nabla_+ - g_Y^2 |\Phi|^2\right)\psi_{+R} + \sqrt{2}g_Y \left(\nabla_+\Phi\right)^* \psi_{-R} = 0, \\
&\left(\partial^2 + 2\nabla_+\nabla_- - g_Y^2 |\Phi|^2\right)\psi_{-R} + \sqrt{2}g_Y \nabla_+\Phi\, \psi_{+R} = 0,
\end{aligned}
\tag{3.3.43}
$$

where Φ represents again the background of the 6D scalar, namely the third line of (3.3.12), and the covariant derivatives are evaluated with the background metric and background gauge field given by the first and the second line of (3.3.12). These covariant derivatives are in the $\pm$ basis defined by (3.3.20) and it includes the modified spin connection when it acts on spinors:

$$
\nabla_\alpha \psi_{\pm R} = e_\alpha^m(y,\eta)\left(\partial_m \pm \omega_m \frac{1}{2} + ie_\pm A_m\right)\psi_{\pm R},
\tag{3.3.44}
$$

$$\nabla_\alpha \psi_{\pm L} = e^m_\alpha(y,\eta)\left(\partial_m \mp w_m\frac{1}{2} + ie_\pm A_m\right)\psi_{\pm L}, \qquad (3.3.45)$$

where $w_\theta = 0$, $w_\varphi \equiv w^+_{\varphi\ +}$ is given in equation (B.1.13) and the value of the charges $e_\pm$ and the iso-helicities[12] of the fermions are given at the end of Subsection 3.2.1. There we give also the fermionic massless spectrum for $\eta = 0$: an $SU(2)$ singlet from ψ_{+L} and an $SU(2)$ triplet from ψ_{-R}.

From (3.3.43) it's clear that the left handed sector does not present mixing terms of the order $\eta^{1/2}$ but only of the order η. Therefore the calculation of the squared mass M^2_F of the light fermion coming from ψ_{+L} is quite easy. The result is

$$M^2_F = \frac{3g^2_Y}{16\pi a^2}\frac{-\mu^2}{\lambda_H - \lambda_G}. \qquad (3.3.46)$$

Instead the evaluation of the right-handed spectrum is complicated by the presence of mixing terms of the order $\eta^{1/2}$, as in the scalar sector. Therefore we use the perturbation theory of quantum mechanics also in the fermion right-handed sector. Formally we can write the eigenvalue equation for the mass squared operator $\mathcal{O}$ acting in the right-handed sector as follows

$$\mathcal{O}F_R = M^2 F_R, \qquad (3.3.47)$$

where F_R is an array which includes both the right-handed fermions; we choose

$$F_R = \begin{pmatrix} \psi_{+R} \\ \psi_{-R} \end{pmatrix}. \qquad (3.3.48)$$

One can easily compute $\mathcal{O}$ acting on F_R by performing the substitution $\partial^2 \to M^2$ in the last two equations of (3.3.43). Then we can proceed as in the scalar spectrum, performing the decomposition (3.3.38). However, in this case the matrix M^2_{ij} in (3.3.39) is a 3×3 matrix as the number of zero modes for $\eta = 0$ in the right-handed sector is 3. Like in the scalar spectrum we need only those $|\tilde{i} >$ vectors made of harmonics with $l \leq 2$, because the operator $\mathcal{O}_1$ modifies the integration measure just by a factor proportional to the harmonics $\mathcal{D}^{(1)}$. In Appendix B.3 we give an expression for the $|i >, i = 1, -1, 0$, vectors, for the $|\tilde{i} >$ vectors and the $M^2_{\tilde{i}}$ eigenvalues for the relevant values of l: $l = 1, 2$. Here we give the final result: the right-handed low energy

---

[12]For $\eta \neq 0$ we adopt the same harmonic expansion as in the $\eta = 0$ case; this gives the correct result for the fermionic masses squared at the order η.

spectrum has a pair of massless right-handed fermions as in the 4D effective theory, which have opposite charge under the residual $U(1)$ symmetry, and a massive right-handed fermion with the same squared mass given in (3.3.46). This right-handed fermion together with the massive left-handed fermion form a massive Dirac spinor with mass M_F.

Also in the fermionic sector we note that the heavy modes contribution is needed in order that the effective theory reproduces the correct 6D result; this sentence is evident if one compares the effective theory prediction (3.2.47) with the correct result (3.3.46).

3.4 Conclusions and Outlook of Part I

The principal result of Chapter 2 and 3 is that the contribution of the heavy KK modes to the effective 4D action is necessary in order to reproduce the correct D-dimensional predictions concerning the light KK modes. We have calculated such a contribution for a class of scalar theories in Chapter 2. However, this result holds in a more general framework. In order to show this, in this chapter we have studied a 6D gauge and gravitational theory which involves a complex scalar and, possibly, fermions. In particular we have considered the compactification over S^2, for a particular value of the monopole number $(n = 2)$, and the construction of a 4D $SU(2) \times U(1)$ effective theory. The latter contains a scalar triplet of $SU(2)$ which, through an Higgs mechanism, gives masses to the vector, scalar and fermion fields. An explicit expressions for these masses and for the VEV of the scalar triplet was found at the leading order in the small mass ratio μ/M, where M is the lightest heavy mass. On the other hand, for $n = 2$, we found a simple perturbative solution of the fundamental 6D EOMs with the same symmetry of the 4D effective theory in the broken phase. This solution presents a deformation of the internal space S^2 to an ellipsoid, which has isometry group $U(1)$ instead of $SU(2)$. Moreover we computed the corresponding vector, scalar and fermion spectrum with quantum mechanics perturbation theory technique. We have demonstrated by direct calculation that these quantities, computed in the 6D approach, are equal to the corresponding predictions of the 4D effective theory only if the contribution of the heavy KK modes are taken into account. In Table 3.1 we give the spectrum predicted by the 4D effective theory for $c_i = 0$, namely, without heavy KK mode contribu-

Squared Mass	4D Effective Theory	6D Theory
M_V^2	$\frac{3e^2}{8\pi a^2}\frac{-\mu^2}{\lambda_H}$	$\frac{3e^2}{8\pi a^2}\frac{-\mu^2}{\lambda_H-\lambda_G}$
$M_{V\pm}^2$	$\frac{9e^2}{16\pi a^2}\frac{-\mu^2}{\lambda_H}$	$\frac{9e^2}{16\pi a^2}\frac{-\mu^2}{\lambda_H-\lambda_G}$
M_S^2	$-2\mu^2$	$-2\mu^2$
$M_{S\pm}^2$	$-\mu^2$	$-\mu^2\frac{\lambda_H+\lambda_G}{\lambda_H-\lambda_G}$
M_F^2	$\frac{3g_Y^2}{16\pi a^2}\frac{-\mu^2}{\lambda_H}$	$\frac{3g_Y^2}{16\pi a^2}\frac{-\mu^2}{\lambda_H-\lambda_G}$
$M_{F\pm}^2$	0	0

Table 3.1: The spectra predicted by the 4D effective theory without heavy modes contribution ($c_i = 0$) and by the 6D theory.

tion, and the low energy spectrum predicted by the 6D theory for the stable ($\mu^2 < 0$) solution, that we gave in the text. We observe that ratios of masses which involve only vector and fermion excitations are correctly predicted by the 4D effective theory even without the heavy KK mode contribution. But the ratios of masses which involve at least one scalar mode are not correctly predicted and the error is measured by λ_G/λ_H, where λ_G and λ_H are defined in equations (3.2.35). We can roughly estimate the magnitude of this disagreement: if we require g_1 and g_2 in (3.2.37) to be of the order of 1 and we consider also the relation between κ and the 4D Planck length κ_4

$$\frac{4\pi a^2}{\kappa^2} = \frac{1}{\kappa_4^2}, \tag{3.4.1}$$

we get that $\sqrt{\kappa}$, e and a are all of the order of κ_4. So roughly speaking the condition $\lambda_G/\lambda_H \ll 1$ becomes $\lambda_H \gg 1$, which is a strong coupling regime. Therefore we can't probably neglect the heavy KK mode contribution and believe in the perturbation theory of quantum field theory at the same time.

Finally we note that there is a value of c_1 and c_2 ($c_1 = -1/3$, $c_2 = 1$) such that the effective theory VEV and vector, scalar and fermion spectrum turn out to be correct, namely, they are equal to the corresponding quantities given in Section 3.3. This is a sign of the equivalence between the geometrical approach, which involves the deformed internal space geometry, to the spontaneous symmetry breaking and the Higgs mechanism in the 4D effective theory. In particular the heavy KK mode contribution can be interpreted in a geometrical way as the internal space deformation of the 6D solution: in fact if we put $\beta = 0$ but we keep $\alpha \neq 0$ in (3.3.12), which corresponds to neglecting the S^2 deformation, we get exactly the VEV and the spectrum predicted by the 4D effective theory without heavy KK modes contribution.

Possible applications can be its extension to the case which resembles more the standard electro-weak theory. The latter could be for instance the 6D gauge and gravitational theory presented in this chapter, compactified over S^2 but with monopole number $n = 1$; in this case we have in fact an Higgs doublet in the 4D effective theory. Other interesting applications could be models without fundamental scalars, which, in some sense, geometrize the Higgs mechanism or the context of supersymmetric version of 6D gauge and gravitational theories. Such supersymmetric theories have been investigated in connection with attempts to find a solution to the cosmological dark energy problem, a summary of which can be found in [55] and in Section 3 of Chapter 4.

After the publication of the author's Ph.D. thesis [56], it was found that the heavy mode contribution, although necessary, is sometimes not sufficient for a proper description with the 4D effective theory approach. This occurs for instant when a gauge field acquires a non-trivial profile along the extra dimensions because of a Higgs mechanism and fermions with different 4D chiralities are localized on different branes (where the gauge field has different values). If the 4D effective theory is contructed before the higher dimensional Higgs mechanism and the spontaneous symmetry breaking is triggered only in 4D, the resulting theory may be vector-like (that is there is no difference between right-handed and left-handed fermions); if instead the 4D effective theory is obtained taking into account the spontaneous symmetry breaking from the very beginning one has a chiral spectrum. This point is described in details in Ref. [57] where an explicit example is provided. In Ref. [57] fermions with different chiralities are localized on different branes by using a method similar (although slightly more sophisticated) to that described in Subsection 1.2.2

Part II: 6D Supergravity

In the second part of this book we consider supersymmetric and non-Abelian extensions of the 6D Einstein-Maxwell-Scalar model that we discussed in Chapter 3. The discussion of chapter 3 was motivated by a theoretical question, concerning the role of heavy modes, with masses of the order of the Planck scale, in the low energy dynamics. Here we want to discuss 6D supergravity from both a theoretical and phenomenological point of view. We know that models in six dimensions are relevant for several reasons; for example the attempt to solve the hierarchy problem (between the electroweak and the Planck scale) by means of the ADD scenario is phenomenologically viable but also falsifiable in 6D, because it is subjected to tests of gravity at sub-millimeter scales, as we discussed in Section 1.3. On the other hand also supersymmetry has several motivations, in particular of the theoretical type. One of them is the fact that superstring theories, the only attempt to unify fundamental interactions including gravity, are supersymmetric; another motivation for supersymmetry is the possibility to solve the hierarchy problem in a supersymmetric framework. Of course this does not mean that one has to choose among LED and supersymmetric theories to address the hierarchy problem: the LED scenario can play a role in addition to supersymmetry, rather than in competition with it. The aim of Part II is to study the implications of 6D supersymmetric models including gravitational interactions and therefore we will deal necessarily with supergravity.

This part contains two chapters. In Chapter 4 we will review the general features of 6D supergravities, focusing on the *minimal gauged* supergravity. In particular we shall discuss the vacua of such models, which *spontaneously compactify* from 6D to 4D and share many properties with realistic string compactifications; moreover we will illustrate the so called *supersymmetric large extra dimensions* scenario in which one *can hope* to solve the cosmological constant problem through a *self tuning* mechanism. However, the embedding of 6D supergravity in the ADD scenario needs the appearance of 3-branes where the low energy degrees of freedom physically localize. So we will review also singular *3-brane solutions* of such models which has to be interpreted as backgrounds around which physical degrees of freedom fluctuate. Indeed in Chapter 5, which report results from [21], we shall study perturbations around such 3-brane solutions, in particular focusing on *axisymmetric solutions*. These solutions will turn out to have conical defects, which we have discussed in Section 1.6. Our main interest will be the gauge field and fermion sectors which can contain SM fields and the effect of the warping and the deficit angles on the KK towers. Moreover, in Appendix C.1 we shall also perform a stability analysis for the only one known maximally symmetric solution, in the presently known anomaly-free models. Finally, in the rest of Appendix C, we will discuss some technical aspects concerning the gauge field and fermion sector.

Chapter 4

General Features

In this review chapter we focus on the minimal supersymmetric version of 6D supergravities, in which we have the minimum number of super-charges in six dimensions. But we consider the possibility of gauging a subgroup of the R-symmetry group which rotates the supercharges; this type of models ([48], [58]-[71]) are called gauged supergravities and they have attracted much interest over the years for several reasons. A reason motivating such models is that the flat 6D space-time *is not* a solution of the corresponding equations of motion (EOM) and the most symmetric solution is $(Minkowski)_4 \times S^2$, which has been shown recently to be the *unique* maximally symmetric solution of such models [63]. This phenomenon of *spontaneous compactification* is a good property which is not shared by 10D and 11D supergravities, as the low energy limit of the superstring theories or the *M-theory*; indeed the most symmetric ground state solutions in all of the higher dimensional supergravities are the flat 10D manifolds and the pp waves. Moreover 6D gauged supergravity compactifications share some properties with superstring realistic compactifications [62], in particular they can give rise to chiral fermions in 4D. Futhermore, like in string theory, the requirement of anomaly freedom is a strong guiding principle to construct consistent models. Indeed the minimal version of such gauged supergravity, *the Salam-Sezgin model* [48], suffers from the breakdown of local symmetries due to the presence of gravitational, gauge and mixed anomalies, which render this model inconsistent at the quantum level [72]; but it can be transformed in an anomaly free model by choosing the gauge group and the supermultiplet in a suitable way

[59, 67, 68, 70]. Recently such 6D supergravities have been proposed as possible frameworks in which one *can hope* to solve the cosmological constant problem. A reason is that, if one chooses large extra dimensions, in 6D the corresponding KK mass scale (to the fourth power) is of the order of the observed vacuum energy density. This numerical coincidence gives hope to get the correct cosmological constant including both classical and quantum contributions, through a mechanism of *self tuning* of the cosmological constant. Such scenario is called *supersymmetric large extra dimensions* (SLED) scenario and it has been studied in recent works[1] [34]-[38]. However, a complete proof of this mechanism has not been found and it is not clear what is the complete effect of the breakdown of supersymmetry in the bulk, which is needed in order to implement such an idea. Moreover, in order the extra dimensions to be so large one should find a mechanism which localizes the low energy degrees of freedom on a *3-brane*, placed on some singularities of the internal space [63, 64, 66, 69].

The aim of this chapter is to review such topics in order to prepare the background for the contents of Chapter 5. The composition of the chapter is as follows. In Section 4.1 we discuss supergravity in diverse dimensions and in Section 4.2 we focus on the minimal 6D *gauged* supergravity by discussing the supermultiplets and the actions for such models and then by describing the presently known anomaly free versions. In Appendix C.1 we perform the stability analysis for an S^2 compactification of these models. In Section 4.3 we describe the SLED scenario in more detail, explaining in particular the self tuning mechanism for the cosmological constant. Finally in Section 4.4 we review the brane solutions of 6D gauged supergravity models.

4.1   Supergravity in Diverse Dimensions

Here we want to discuss briefly general properties of supergravities in diverse dimensions, in order to introduce notations and terminology. For a more complete introduction to such a topic see for instance Ref. [73]-[75].

The starting point to construct a supersymmetric model, in particular a supergravity model, is the choice of a *superalgebra* or *super-Poincaré algebra*. The latter includes by definition the generators of

---

[1]For a review on this topic see [55].

D	Spinor	Components
2 mod 8	Maj-Weyl	$2^{D/2-1}$
3,9 mod 8	Maj	$2^{(D-1)/2}$
4,8 mod 8	Maj or Weyl	$2^{D/2}$
5,7 mod 8	Dirac	$2^{(D+1)/2}$
6 mod 8	Weyl	$2^{D/2}$

Table 4.1: Existence of Weyl, Majorana and Majorana-Weyl spinors in diverse space-time dimensions. Here we give also the number of real components.

the Poincaré group and a set of supercharges, which generates supersymmetry. In order to be consistent such generators have to be spinors, that is objects transforming under the spinorial representation of the Lorentz group $SO(1, D-1)$. The generators in such a representation are given by $\frac{1}{4}[\Gamma^M, \Gamma^N]$, where the matrices Γ^M obey the Clifford algebra:

$$\{\Gamma^M, \Gamma^N\} = 2\eta^{MN}. \qquad (4.1.1)$$

In this way we define a *Dirac spinor*, which has real[2] dimension $2^{[D/2]+1}$, where $[D/2]$ means the integer part of $D/2$, and exists in all space-time dimensions. However, when D is even the Dirac spinor is a reducible representation because one can define a chirality matrix, which is given by the product of all the matrices Γ^M and commutes with all the generators $\frac{1}{4}[\Gamma^M, \Gamma^N]$. The eigenvectors of the chirality matrix are called Weyl spinors. Moreover for some particular value of D we can impose a *reality condition* on spinors and define *Majorana spinors*. In Table 4.1 we summarize the existence of Weyl, Majorana and Majorana-Weyl spinors in diverse space-time dimensions. Besides generators of the Poincaré group and supercharges, the superalgebra can include also a set of gauge bosonic generators.

The requirement that the action functional is invariant under local supersymmetry implies the presence of gravitational interactions, due to the generators of translations in the superalgebra. So local supersymmetry and supergravity are equivalent names for such theories. Therefore supergravity has to contain a metric tensor G_{MN} which turns to be necessarily associated to a ψ_M called gravitino, which is labeled by both spinor and vector indices. In Table 4.2 we give the on shell degrees of freedom of the metric and the gravitino as function of

---

[2]In this book we refer always to the real spinor dimension.

Field	Spin	On-shell d.o.f.
G_{MN}	2	$(D-2)(D-1)/2 - 1$
ψ_M	3/2	$(D-3)\mathcal{I}/2$

Table 4.2: On shell degrees of freedom of the metric G_{MN} and the gravitino ψ_M, which always appear in the *supergravity multiplet*. The integer $\mathcal{I}$ represents the number of components of the irreducible spinorial representation.

D.

Both for local and global supersymmetry the total number of supercharges must be a multiple N of the components of the irreducible spinorial representation of the Lorentz group. Since there are no consistent quantum field theories including fields whose spin is greater than 2 and no non-gravitational field theories including fields with spin greater than one, one can get a constraint on N. In particular there are no consistent 4D field theories with N>8 and no consistent 4D field theories without gravity with N>4. Since the dimension of the irreducible spinorial representation depends on D the maximum value of N, which gives rise to a consistent theory, depends on D as well. For instance 10D supergravity has at most N=2 and 11D supegravity has necessarily N=1.

4.2   The 6D Case

Now we focus on 6D supergravity which is one of the main topic of this book. Contrary to the 4D case, supercharges with positive chirality and with negative chirality do not give rise to equivalent models. Therefore we introduce the notation

$$N = (N_+, N_-), \qquad (4.2.2)$$

where $N_\pm$ is the number of supercharges with positive (negative) chirality. If $N_+ \neq N_-$ the corresponding supergravity is called chiral. There exist several possibilities: $N = (1,0)$, $N = (1,1)$, $N = (2,0)$, $N = (2,2)$ and $N = (4,0)$. Although in this book we are interested in the minimal version $N = (1,0)$ of such models, in the following we shall describe briefly all the possibilities for the sake of completeness.

We start with the **chiral** $N = (1,0)$ **supergravity**, which has a number of supercharges equal to $N = 2$ supersymmetry in 4D. Indeed the R-symmetry group turns out to be $Sp(1) = SU(2)$ and henceforth it will be denoted by $Sp(1)_R$. The supergravity multiplet consists of

$$\left(G_{MN}, \psi_M^j, B_{MN}^+\right),\tag{4.2.3}$$

where j takes value in the fundamental of $Sp(1)_R$, ψ_M^j has a positive chirality, that is[3]

$$\Gamma^7 \psi_M^j = \psi_M^j,\tag{4.2.4}$$

and B_{MN}^+ represents a *self-dual* field strength. However, it is well known that field theories with self-dual field strength do not admit a manifestly Lorentz invariant action formulation, in dimensions 2 mod 4 [76]. This problem can be avoided if one combines the multiplet (4.2.3) with a *tensor multiplet*:

$$\left(B_{MN}^-, \chi^j, \sigma\right),\tag{4.2.5}$$

where B_{MN}^- is an *antiself-dual* field strength, the spinor χ^j is called tensorino and it satisfies

$$\Gamma^7 \chi^j = -\chi^j\tag{4.2.6}$$

and σ is a real scalar field, which is called dilaton. Now one can form a generic field strength $B_{MN} = B_{MN}^+ + B_{MN}^-$ and give a manifestly Lorentz invariant action formulation. We shall refer to B_{MN} as a Kalb-Ramond field.

Moreover there exist also Yang-Mills multiplets, corresponding to a gauge group $\mathcal{G}$

$$\left(\mathcal{A}_M^I, \lambda^{Ij}\right),\tag{4.2.7}$$

where I is a Lie algebra index, $\mathcal{A}_M^I$ are the gauge fields and λ^{Ij} the gauginos, which satisfy

$$\Gamma^7 \lambda^{Ij} = \lambda^{Ij}.\tag{4.2.8}$$

Finally we can introduce also hypermultiplets:

$$\left(\psi^a, \phi^\alpha\right),\tag{4.2.9}$$

where ψ^a, $a = 1, ..., 2n_H$, called hyperinos, satisfy

$$\Gamma^7 \psi^a = -\psi^a\tag{4.2.10}$$

---

[3]In Appendix A we give our conventions on the gamma matrices in the 6D case.

$G_s/H_s \times Sp(1)_R$	H_s-representation of ψ^a
$Sp(n,1)/Sp(n) \times Sp(1)_R$	$\mathbf{2n}$
$SU(n,2)/SU(n) \times U(1) \times Sp(1)_R$	$\mathbf{n_q} + \mathbf{n_{-q}}$
$SO(n,4)/SO(n) \times SO(3) \times Sp(1)_R$	$(\mathbf{n}, \mathbf{2})$
$E_8/E_7 \times Sp(1)_R$	$\mathbf{56}$
$E_7/SO(12) \times Sp(1)_R$	$\mathbf{32}$
$E_6/SU(6) \times Sp(1)_R$	$\mathbf{20}$
$F_4/Sp(3) \times Sp(1)_R$	$\mathbf{14}$
$G_2/Sp(1) \times Sp(1)_R$	$\mathbf{4}$

Table 4.3: Quaternionic symmetric spaces parametrized by the hyperscalars. These are coset space of the form $G_s/H_s \times Sp(1)_R$, where G_s is a group and H_s a subgroup of G_s. We give also the corresponding hyperinos representation.

and ϕ^α, $\alpha = 1, ..., 4n_H$ are called hyperscalars. The latter parametrize a manifold which is non-compact and quaternionic: a quaternionic manifold is a Riemannian manifols with holonomy group[4] contained in $Sp(n) \times Sp(1)$. In Table 4.3 we give the quaternionic symmetric spaces parametrized by hyperscalars for this type of supergravity.

The gauge group $\mathcal{G}$ can be taken to be a direct product of an arbitrary gauge group, which does not act on hypermultiplets, times a group H_s, defined in Table 4.3, or subgroup of H_s, which do act on hypermultiplets. We observe that hypermultiplets are always neutral with respect to $Sp(1)_R$, which can be either gauged or ungauged. In Section 4.2.1 we will discuss more the $N = (1,0)$ model gauging part of the isometry group of the hyperscalar manifold and giving the explicit expression for the action functional in this model.

We continue our analysis of 6D supergravities considering the **non chiral** $N = (1,1)$ **model**. This is an $SU(2)$ gauged supergravity consisting of

$$\left(G_{MN}, \psi_{Mi}, B_{MN}, A_M, A^j_{Mi}, \lambda_i, \phi\right) \qquad (4.2.11)$$

---

[4]On a Riemannian manifold, tangent vectors can be moved along a path by parallel transport, which preserves vector addition and scalar multiplication. So a closed loop at a base point p, gives rise to a invertible linear map of the tangent space of the manifold in p. It is possible to compose closed loops by following one after the other, and to invert them by going backwards. Hence, the set of linear transformations arising from parallel transport along closed loops is a group, called the holonomy group.

coupled to a vector multiplet

$$\left(B_M, \chi_i, A_i^j, \xi\right). \tag{4.2.12}$$

Minimal versions of these models has no stable maximally symmetric ground state, that is Minkowski, de Sitter or anti de Sitter space. However, it is possible to generalize this framework, by introducing a mass parameter for the 2-form tensor B_{MN}, in a way that there exist maximally symmetric solutions.

Another possible version is the **chiral** $N = (2,0)$ **supergravity**. The supergravity multiplet consist of

$$\left(G_{MN}, \psi_M^k, B_{MN}^{+kl}\right), \tag{4.2.13}$$

where k, l label **4** of $USp(4)$ and the 2-form tensor field is in the **5** of $USp(4)$. This supergravity admits couplings to a tensor multiplet which contains

$$\left(B_{MN}^-, \lambda^k, \phi^{kl},\right). \tag{4.2.14}$$

One can introduce an arbitrary number n of tensor multiplets and the scalars parametrize $SO(n,5)/SO(n) \times SO(5)$ and the $(n+5)$ 2-form fields trasforms as $(n+5)$ of $SO(n,5)$.

Moreover a **non chiral** $N = (2,2)$ and **two chiral** $N = (4,0)$ supergravities exist. The chiral versions have exotic field contents which does not include a graviton.

4.2.1 The $N = (1,0)$ Gauged Supergravity

Now we want to discuss in more detail the action formulation of the $N = (1,0)$ model, gauging the isometry group of the hyperscalar manifold. This derivation has been done in [58] but here we give the explicit expression for the action as it is relevant for the developments of Chapter 5.

As a first step we choose one quaternionic manifold targed by the hyperscalars. The most common choice is $Sp(n,1)/Sp(n) \times Sp(1)_R$, appearing in the first row of Table 4.3. This is also the choice of Ref. [58]. We remind that hypermultiplets transform non-trivially under the isometry group $Sp(n,1)$ of such manifold. Indeed the hypermultiplets turn out to be in the representation **2n** of $Sp(n)$. Here we consider the gauging of the complete $Sp(n) \times Sp(1)_R$ but, after giving the action in this case, we will explain what changes if we gauge only a subgroup of such group. Many geometric properties of such a manifold can be specified by introducing a representative L^{aj} of the coset. This is a map from the coset to the group $Sp(n,1)$ and therefore depend on ϕ^{α}. The *Maurer Cartan* 1-form decomposes as [5]

$$L^{-1}\partial_{\alpha}L = A_{\alpha}^{i}T^{i} + A_{\alpha}^{\hat{I}}T^{\hat{I}} + V_{\alpha}^{aj}T_{aj}, \tag{4.2.15}$$

where T^{i}, $i = 1, 2, 3$, and $T^{\hat{I}}$, $\hat{I} = 1, ..., n(2n + 1)$, are the anti-hermitian generators of $Sp(1)_{R}$ and $Sp(n)$ respectively, while T_{aj} are the anti-hermitian coset generators. The objects A_{α}^{i} and $A_{\alpha}^{\hat{I}}$ transforms as $Sp(1)_{R}$ and $Sp(n)$ connection, respectively, while V_{α}^{aj} transforms homogeneously under the induced local tangent space transformations. Thus A_{α}^{i} and $A_{\alpha}^{\hat{I}}$ can be used in the definition of the $Sp(n) \times Sp(1)_{R}$ covariant derivatives and V_{α}^{aj} can be used as a frame on the coset space and it is covariantly constant with respect to the composite $Sp(n) \times Sp(1)_{R}$ connections and the Christoffel connection defined on $Sp(n,1)/Sp(n) \times Sp(1)_{R}$ in the usual way. As a result, the $Sp(n) \times Sp(1)_{R}$ connections can be expressed in terms of the frames V_{α}^{aj}. The gauging of $Sp(n) \times Sp(1)_{R}$ can be implemented by using these objects. For example, the covariant derivative of the hyperinos can be written as follows

$$\nabla_{M}\psi = \left(\partial_{M} + \frac{1}{8}\omega_{M}^{[A,B]}[\Gamma_{A}, \Gamma_{B}] + \partial_{M}\phi^{\alpha}A_{\alpha}^{\hat{I}}T^{\hat{I}}\right)\psi, \tag{4.2.16}$$

where $\omega_{M}^{[A,B]}$ is the Lorentz connection defined in Appendix A. Moreover the covariant derivative of hyperscalars is

$$\nabla_{M}\phi^{\alpha} = \partial_{M}\phi^{\alpha} - A_{M}^{\hat{I}}\xi^{\alpha\hat{I}} - A_{M}^{i}\xi^{\alpha i}, \tag{4.2.17}$$

where the $\xi^{\alpha I}$ are defined by

$$\xi^{\alpha I} = \left(T^{I}\phi\right)^{\alpha}. \tag{4.2.18}$$

The elements discussed so far are sufficient to derive an action functional (up to quartic fermionic terms) for $N = (1,0)$ gauged supergravity:

$$
\begin{aligned}
S \;=\; \int d^6 X \sqrt{-G} \Bigg[& \frac{1}{\kappa^2} R - \frac{1}{4} \partial_M \sigma \partial^M \sigma - \frac{\kappa^2}{48} e^{\kappa\sigma} G_{MNR} G^{MNR} \\
& - \frac{1}{4} e^{\kappa\sigma/2} \left(\frac{1}{\hat{g}^2} \hat{F}^2 + \frac{1}{g_1^2} F_1^2 \right) - g_{\alpha\beta}(\phi) \nabla_M \phi^\alpha \nabla^M \phi^\beta \\
& - \frac{8}{\kappa^4} e^{-\kappa\sigma/2} C^{i I} C^{i I} + \frac{1}{2} \overline{\psi}_M \Gamma^{MNR} \nabla_N \psi_R \\
& + \frac{1}{2} \overline{\chi} \Gamma^M \nabla_M \chi + \frac{1}{2} \overline{\lambda} \Gamma^M \nabla_M \lambda + \frac{1}{2} \overline{\psi}_a \Gamma^M \nabla_M \psi^a \\
& + \frac{\kappa}{4} \overline{\chi} \Gamma^N \Gamma^M \psi_N \partial_M \sigma - \frac{\kappa}{2} \overline{\psi}_M^j \Gamma^N \Gamma^M \psi^a \nabla_N \phi^\alpha V_{\alpha a j} \\
& + \frac{\kappa^2}{96} e^{\kappa\sigma/2} G_{MNR} \left(\overline{\psi}^L \Gamma_{[L} \Gamma^{MNR} \Gamma_{T]} \psi^T + 2 \overline{\psi}_L \Gamma^{MNR} \Gamma^L \chi \right. \\
& \left. - \overline{\chi} \Gamma^{MNR} \chi + \overline{\lambda} \Gamma^{MNR} \lambda + \overline{\psi}_a \Gamma^{MNR} \psi^a \right) \\
& - \frac{\kappa}{4\sqrt{2}\hat{g}} e^{\kappa\sigma/4} \hat{F}_{MN}^{\hat{I}} \left(\overline{\psi}_L \Gamma^{MN} \Gamma^L \lambda^{\hat{I}} + \overline{\chi} \Gamma^{MN} \lambda^{\hat{I}} \right) \\
& - \frac{\kappa}{4\sqrt{2}g_1} e^{\kappa\sigma/4} F_{1MN}^i \left(\overline{\psi}_L \Gamma^{MN} \Gamma^L \lambda^i + \overline{\chi} \Gamma^{MN} \lambda^i \right) \\
& + \frac{\kappa}{2\sqrt{2}} e^{-\kappa\sigma/4} \left(\overline{\psi}_M \Gamma^M T^i \lambda C^i - \overline{\chi} T^i \lambda C^i \right. \\
& \left. - 2 \overline{\psi}^a \lambda^j V_{\alpha a j} \tilde{\xi}^\alpha \right) \Bigg],
\end{aligned}
\tag{4.2.19}
$$

where $(\hat{g}, \hat{F})$ and (g_1, F_1) are the gauge constants and the field strengths of $Sp(n)$ and $Sp(1)_R$ respectively, the 3-form G_3 is defined by

$$
G_3 = dB_2 + \frac{1}{g^2} \left(A \wedge F - \frac{2}{3} A \wedge A \wedge A \right),
\tag{4.2.20}
$$

where g is a generic gauge constant, moreover

$$
\tilde{\xi}^{\alpha I} \equiv \left(\hat{g} \xi^{\alpha \hat{I}}, g_1 \xi^{\alpha i} \right)
\tag{4.2.21}
$$

and the C-functions are defined by

$$
C^{i\hat{I}} = \hat{g} A_\alpha^i \xi^{\alpha \hat{I}}, C^{ik} = g_1 \left(A_\alpha^i \xi^{\alpha k} - \delta^{ik} \right).
\tag{4.2.22}
$$

In the second line of (4.2.19) a trace over the gauge group generators is understood. An important feature of this action is it has a positive definite potential.

As we have already mentioned, the action in (4.2.19) corresponds to the gauging of the complete $Sp(n) \times Sp(1)_R$ group. If we want to gauge just a subgroup, for example $E_7 \times U(1)_R$, only the gauge field strengths, the gauginos and the C-functions of such a subgroup will appear in the action. Moreover one can also introduce additional vector multiplets containing the gauge fields of an additional *group factor* which is not a subgroup of $Sp(n) \times Sp(1)_R$. The hypermultilpets are neutral with respect to the additional gauge interactions. In this case one has to add the corresponding Yang-Mills fields and gauginos in the action as in the explicit example treated in [59], in particular the additional gauginos will appear only in line 4 and 8 of (4.2.19).

The action and the field content that we have discussed so far can be generalized to include all the quartic fermionic terms [60] and a generic number n_T of tensor multiplets [61], [77], [78].

The elements that we discussed in this section are quite general but in the next subsection we shall discuss some particular examples.

4.2.2   Salam-Sezgin Model and Anomaly-Free Versions

Now we turn to describe the simplest example of 6D $N = (1,0)$ gauged supergravity: the so called Salam-Sezgin model [48]. In the bosonic sector of this model we have only the metric, the Kalb-Ramond field, the dilaton and the gauge field associated to a subgroup $U(1)_R$ of $Sp(1)_R$. This can be considered as a supersymmetrization of the 6D Einstein-Maxwell model of Chapter 3 as the complete gauge group is $U(1)_R$. Moreover in the fermionic sector we have the $U(1)_R$ gaugino in the vector multiplet, the gravitino in the supergravity multiplet and the tensorino in the tensor multiplet. The complete action for this system can be obtained by putting equal to zero in (4.2.19) all the fields that we have not mentioned. In particular there are no hyperscalars in the bosonic sector. Therefore the bosonic part S_B of the action is

$$
S_B = \int d^6 X \sqrt{-G} \left[\frac{1}{\kappa^2} R - \frac{1}{4} \partial_M \sigma \partial^M \sigma - \frac{\kappa^2}{48} e^{\kappa\sigma} G_{MNR} G^{MNR} \right.
$$
$$
\left. - \frac{1}{4g_1^2} e^{\kappa\sigma/2} F_1^2 - \frac{8g_1^2}{\kappa^4} e^{-\kappa\sigma/2} \right], \qquad (4.2.23)
$$

and the corresponding EOM are

$$\frac{1}{\kappa^2}R_{MN} = \frac{1}{2g_1^2}e^{\kappa\sigma/2}F_{1MP}F_{1N}{}^{P} + \frac{1}{4}\partial_M\sigma\partial_N\sigma$$
$$+\frac{\kappa^2}{16}e^{\kappa\sigma}G_{MPQ}G_N{}^{PQ} - \frac{1}{4\kappa}G_{MN}\nabla^2\sigma,$$

$$\frac{1}{\kappa}\nabla^2\sigma = \frac{1}{4g_1^2}e^{\kappa\sigma/2}F_1^2 + \frac{\kappa^2}{24}e^{\kappa\sigma}G_{MNP}G^{MNP} - \frac{8g_1^2}{\kappa^4}e^{-\kappa\sigma/2},$$

$$\nabla_M\left(e^{\kappa\sigma/2}F_1^{MN}\right) = \frac{\kappa^2}{4}e^{\kappa\sigma}G^{NPQ}F_{1PQ},$$

$$\nabla_M\left(e^{\kappa\sigma}G^{MNP}\right) = 0. \tag{4.2.24}$$

We look now for maximally symmetric solutions of this model which preserve a 4D Poincaré invariance. To this end we observe that 4D Poincaré invariance implies $G_{MNR} = 0$ at the background level. By using this constraint, the EOM (4.2.24) reduce to

$$\frac{1}{\kappa^2}R_{MN} = \frac{1}{2g_1^2}e^{\kappa\sigma/2}F_{1MP}F_{1N}{}^{P} + \frac{1}{4}\partial_M\sigma\partial_N\sigma - \frac{1}{4\kappa}G_{MN}\nabla^2\sigma,$$

$$\frac{1}{\kappa}\nabla^2\sigma = \frac{1}{4g_1^2}e^{\kappa\sigma/2}F_1^2 - \frac{8g_1^2}{\kappa^4}e^{-\kappa\sigma/2},$$

$$\nabla_M\left(e^{\kappa\sigma/2}F^{MN}\right) = 0. \tag{4.2.25}$$

Since we are looking for a maximally symmetric solution we suppose σ constant and inserting this ansatz in (4.2.25) we get, after some manipulations,

$$R = \frac{16g_1^2}{\kappa^2}e^{-\kappa\sigma/2}, \tag{4.2.26}$$

that is the Ricci scalar is a non vanishing constant. Therefore the 6D Minkowski space is not a solution of the Einstein equation appearing in (4.2.25). Since S^2 is the only orientable 2D manifold with positive constant curvature an obvious solution of (4.2.26) is $(Minkowski)_4 \times S^2$. Even if a priori we cannot exclude the presence of a *warp factor*, in Ref. [63] it was proved that this is actually the unique maximally symmetric smooth solution for the metric. In particular the 6D de Sitter and Anti de Sitter spaces are not solutions of the EOM. Requiring the other fields to have the same symmetry as the metric one gets the complete *Salam-Sezgin background*:

$$ds^2 = \eta_{\mu\nu}dx^\mu dx^\nu + a^2\left(d\theta^2 + \sin^2\theta d\varphi^2\right),$$
$$\mathcal{A} = \frac{n}{2}(\cos\theta \pm 1)d\varphi,$$
$$\sigma = \sigma_0 = constant, \quad G_{MNR} = 0. \tag{4.2.27}$$

This is a solution of (4.2.25) if

$$\frac{1}{\kappa^2} = \frac{n^2 e^{\kappa\sigma_0/2}}{8g_1^2 a^2}, \qquad \frac{1}{\kappa^4} = \frac{n^2 e^{\kappa\sigma_0}}{64g_1^4 a^4}. \tag{4.2.28}$$

We observe that first and second line of (4.2.27) correspond exactly to equations (3.2.1) and (3.2.2) of Chapter 3, that is the monopole configuration over S^2. Dirac quantization condition still holds, that is n has to be an integer. However, supersymmetry of the present model and the presence of the dilaton convert constraints (3.2.4) into (4.2.28). An interesting consequence is that now the monopole number n must satisfy

$$n = \pm 1, \tag{4.2.29}$$

which can be proved squaring first constraint in (4.2.28) and then using second constraint in (4.2.28). A detailed analysis of the local supersymmetry transformations shows that background (4.2.27) preserves $1/2$ of the 6D $N = (1,0)$ supersymmetries of this model [48], that is a 4D $N = 1$ supersymmetry.

Since the Salam-Sezgin model is the simplest realization of the 6D $N = (1,0)$ gauged supergravity, it was used as a toy model to study the properties of this class of models and of higher dimensional supergravities. For instance the authors of [62] constructed the 4D $N = 1$ supergravity, which describes the low energy dynamic of the Salam-Sezgin model expanded around the background (4.2.27), in order to provide a simple setting sharing the main properties of realistic string compactifications. Moreover in [63] additional solutions of this model, which present conical and non conical singularities, were found[5].

On the other hand, like most 6D supergravities, the Salam-Sezgin model suffers from the breakdown of local symmetries due to the presence of gravitational, gauged and mixed anomalies. Therefore such a model must be enlarged to include additional supermultiplets. It is

---

[5] We shall describe these solutions in Section 4.4.

interesting that the requirement of anomaly freedom is a strong guiding principle to select consistent models, like in 10D supergravities. So this is another property shared by the 6D $N = (1,0)$ gauged models and higher dimensional supergravities.

Until recently the only one known anomaly free model of this type was the $\mathcal{G} = E_7 \times E_6 \times U(1)_R$ model, where E_7 is a subgroup[6] of the $Sp(n)$ that we discussed in Subsection 4.2.1, E_6 is an additional *group factor* and $U(1)_R$ is a subgroup of $Sp(1)_R$. As we discussed in Subsection 4.2.1 the hypermultiplets are singlets with respect to E_6 and $U(1)_R$ but, in order to cancel the anomalies by means of the *Green-Schwarz mechanism*, they are in representation **912** of E_7. Recently more example of anomaly-free models were found. For instance in [67] a $\mathcal{G} = E_7 \times G_2 \times U(1)_R$ model was proposed, with $E_7 \times G_2 \subset Sp(392)$ and the hypermultiplets in the representation $(\mathbf{56}, \mathbf{14})$ of $E_7 \times G_2$. Moreover in [68] the authors proposed $\mathcal{G} = F_4 \times Sp(9) \times U(1)_R$ with hypermultiplets in the representation $(\mathbf{52}, \mathbf{18})$ of $F_4 \times Sp(9)$. Finally in Ref. [70] a huge number of simple anomaly free models was presented with $\mathcal{G}$ given by products of $U(1)$ and/or $SU(2)$ and particular hyperinos representations.

Therefore non-Abelian extensions of the Salam-Sezgin model, including hypermultiplets, are very interesting because they are needed for the consistency at the quantum level. The gauge group for these models will be $\mathcal{G} = \tilde{\mathcal{G}} \times \mathcal{G}_R$, where $\mathcal{G}_R$ is a subgroup of $Sp(1)_R$ and $\tilde{\mathcal{G}}$ is composed by a subgroup of $Sp(n)$ and in case by an additional gauge group[7]. In the following we shall describe the S^2 compactification of such non-Abelian models [59], [67], with a monopole embedded in the Cartan subalgebra of $\tilde{\mathcal{G}} \times \mathcal{G}_R$. This set up is the only compatible with maximal symmetry.

Like the Salam-Sezgin model the bosonic theory includes the metric, the Kalb-Ramond field and the dilaton but here we have also the gauge fields of the complete gauge group $\mathcal{G} = \tilde{\mathcal{G}} \times \mathcal{G}_R$ and the hyperscalars ϕ^α, in some representation of the subgroup of $Sp(n)$. The

---

[6]In this case $n = 456$.
[7]For instance E_6 in the $E_7 \times E_6 \times U(1)_R$ model.

complete bosonic action for these fields is

$$
S_B = \int d^6 X \sqrt{-G} \left[\frac{1}{\kappa^2} R - \frac{1}{4} \partial_M \sigma \partial^M \sigma - \frac{\kappa^2}{48} e^{\kappa \sigma} G_{MNR} G^{MNR} \right.
$$
$$
- \frac{1}{4} e^{\kappa \sigma / 2} \left(\frac{1}{\tilde{g}^2} \tilde{F}^2 + \frac{1}{g_1^2} F_1^2 \right) - g_{\alpha\beta}(\phi) \nabla_M \phi^\alpha \nabla^M \phi^\beta
$$
$$
\left. - \frac{8}{\kappa^4} e^{-\kappa \sigma / 2} C^{iI} C^{iI} \right], \tag{4.2.30}
$$

The symbols have been defined in Subsection 4.2.1. In [65] the explicit expression for the C-functions in the last line of (4.2.30) for $\mathcal{G} = E_7 \times E_6 \times U(1)_R$ have been given, but the result is applicable to any other model of this type. A remarkable result is that the absolute minimum of the scalar potential is at $\phi^\alpha = 0$ [60, 65]. This property suggest us to set $\phi^\alpha = 0$ at the background level in order to get a stable solution and henceforth we will assume that. Moreover in the following we will focus on the case $\mathcal{G}_R = U(1)_R$. This is true for all the anomaly-free models with a gauge group that contains the gauge group of the SM. In this case the bosonic EOM are

$$
\frac{1}{\kappa^2} R_{MN} = \frac{1}{2} e^{\kappa \sigma / 2} \left(\frac{1}{\tilde{g}^2} \tilde{F}_{MP} \tilde{F}_N^{\ P} + \frac{1}{g_1^2} F_{1MP} F_{1N}^{\ P} \right) + \frac{1}{4} \nabla_M \sigma \nabla_N \sigma
$$
$$
+ \frac{\kappa^2}{16} e^{\kappa \sigma} G_{MPQ} G_N^{\ PQ} - \frac{1}{4\kappa} G_{MN} \nabla^2 \sigma,
$$
$$
\frac{1}{\kappa} \nabla^2 \sigma = \frac{1}{4} e^{\kappa \sigma / 2} \left(\frac{1}{\tilde{g}^2} \tilde{F}^2 + \frac{1}{g_1^2} F_1^2 + \frac{\kappa^2}{24} e^{\kappa \sigma} G_{MNP} G^{MNP} \right)
$$
$$
- \frac{8 g_1^2}{\kappa^4} e^{-\kappa \sigma / 2},
$$
$$
\nabla_M \left(e^{\kappa \sigma / 2} \tilde{F}^{MN} \right) = \frac{\kappa^2}{4} e^{\kappa \sigma} G^{NPQ} \tilde{F}_{PQ},
$$
$$
\nabla_M \left(e^{\kappa \sigma / 2} F_1^{MN} \right) = \frac{\kappa^2}{4} e^{\kappa \sigma} G^{NPQ} F_{1PQ},
$$
$$
\nabla_M \left(e^{\kappa \sigma} G^{MNP} \right) = 0. \tag{4.2.31}
$$

We want to discuss the following maximally symmetric solution

$$
ds^2 = \eta_{\mu\nu} dx^\mu dx^\nu + a^2 \left(d\theta^2 + \sin^2 \theta d\varphi^2 \right),
$$
$$
\mathcal{A} = \frac{n}{2} Q(\cos \theta \pm 1) \quad \sigma = constant,
$$
$$
\sigma = \sigma_0 = constant, \quad G_{MNR} = 0 \tag{4.2.32}
$$

where Q is a generator of a $U(1)$ subgroup of a simple factor of $\mathcal{G}$, satisfying $Tr\left(Q^2\right) = 1$, and n is a real number which takes discrete values because of the Dirac quantization condition. In the following we will denote by $U(1)_M$ the abelian group generated by Q. The configuration (4.2.32) is the trivial generalization of (4.2.27) to a non-Abelian model. The background gauge field in (4.2.32) is a monopole configuration and, from group theory point of view, it breaks $\mathcal{G}$ to some subgroup H, which is generated by the generators of $\mathcal{G}$ which commute with Q. Moreover the equations (4.2.31) implies [59]

$$\frac{1}{\kappa^2} = \frac{n^2 e^{\kappa \sigma_0/2}}{8g^2 a^2}, \qquad \frac{g_1^2}{\kappa^4} = \frac{n^2 e^{\kappa \sigma_0}}{64 g^2 a^4}, \qquad (4.2.33)$$

where g is the gauge constant corresponding to the background gauge field. The constraints (4.2.33) represent the generalization of (4.2.28) and they imply the following equation

$$n^2 = \frac{g^2}{g_1^2}. \qquad (4.2.34)$$

If $g = g_1$ we have $n = \pm 1$, that is Eq. (4.2.29), but this is not needed if we embed $U(1)_M$ in $\tilde{\mathcal{G}}$. An important aspect is that the only monopole embedding which preserves part of the 6D $N = (1,0)$ supersymmetries is Q along the Lie algebra of $U(1)_R$ and in this case we have $g = g_1$. No other embedding ($g \neq g_1$) preserve any residual supersymmetry.

Given the symmetries of the problem, we can expect that the low energy effective gauge group is $\mathcal{H} \times \mathcal{G}_{KK}$, where $\mathcal{H}$ is the subgroup of $\mathcal{G}$ that commutes with $U(1)_M \subset \mathcal{G}$ in which the monopole lies, and is orthogonal to[8] $U(1)_M$; moreover $\mathcal{G}_{KK}$ is the KK gauge group coming from the isometry of the internal space.

An explicit example is Q along the Lie algebra of E_6 in the $E_7 \times E_6 \times U(1)_R$ model [59]. From the purely group theory point of view such an embedding breaks E_6 down to $SO(10) \times U(1)_M$ and leaves $E_7 \times U(1)_R$ unbroken. However, the gauge group of the 4D effective theory contains an additional SU(2) factor, coming from the isometry of S^2. This is a well known mechanism in KK theories and we shall denote this extra group by $SU(2)_{KK}$. Moreover, due to the *Chern-Simons coupling* in supergravity, the $U(1)_M$ gauge field eats

---

[8]Due to the Chern-Simons coupling in supergravity, the $U(1)$ gauge field in the direction of the monopole eats the axion arising from the Kalb-Ramond field and acquires a mass [59, 62].

the axion arising from the Kalb-Ramond field and acquires a mass [59, 62]. So the complete gauge group of the 4D effective theory is $E_7 \times SO(10) \times U(1)_R \times SU(2)_{KK}$. Since the only fermions that interact with the background gauge field[9] are the gauginos of E_6, the latter sector contains the fermionic zero modes. The 4D massless chiral fermions comprise $2|n|$ families of $SO(10)$, in the representation **16** and no antifamilies. They belong to the $|n|$-dimensional irreducible representation of $SU(2)_{KK}$. These fermions are neutral with respect to E_7 but they carry a $U(1)_R$ charge, half the families are positive and the other half are negative. Such **16**-families can be interpreted as leptons and quarks of the SM embedded in a grand unification scenario. Unfortunately the classical stability requires $|n| = 1$ and therefore only 2 families [59, 67, 68]. The bosonic sector of the low energy 4D effective theory contains the graviton and the gauge fields corresponding to $E_7 \times SO(10) \times U(1)_R \times SU(2)_{KK}$. The gauge fields interacting with leptons and quarks are only the gauge fields of $SO(10) \times U(1)_R$ and of $SU(2)_{KK}$ if $|n| > 1$. There is not a complete discussion of the stability issue in the literature. In Appendix C.1 we will give a first step toward this direction following the lines of [59, 67]. In addition to [59, 67] we will include all the present known anomaly free models. The result is that a stable sphere compactification and the embedding of the SM gauge group is possible only in the $E_7 \times E_6 \times U(1)_R$ case.

4.3   Supersymmetric Large Extra Dimensions

So far we have analysed 6D supergravities without any assumptions on the size of the extra dimensions. A physical applications of this framework can be made in the context of LED. As we have mentioned at the beginning of the present chapter, SLED give hope to solve the cosmological constant problem. The aim of this section is to explain this idea in more detail [55] and point out its shortcomings.

In order to discuss the cosmological constant problem, let us consider a generic parameter p which describes a physical quantity and is found to be small when measured in an experiment which is performed at an energy scale μ. We would like to understand this in terms of a

---

[9]There exist no normalizable fermionic zero modes on S^2 satisfying a standard Dirac equation in the absence of coupling to the background monopole field.

microscopic theory which is defined at energy $\Lambda \gg \mu$ and predicts the value of $p(\mu)$ as follows

$$p(\mu) = p(\Lambda) + \delta p(\mu, \Lambda), \qquad (4.3.35)$$

where $p(\Lambda)$ represents the contribution to p due to the parameters in the microscopic theory, and δp represents the contributions to p which are obtained as we integrate out all of the physics in the energy range $\mu < E < \Lambda$. The smallness of $p(\mu)$ can be understood if both $p(\Lambda)$ and δp are small. Although we may not be able to understand why $p(\Lambda)$ is small until we have a correct microscopic theory (up to some energy scale), we should be able to understand why ordinary physics at energies $\mu < E < \Lambda$ do not make $\delta p(\mu, \Lambda)$ unacceptably large. If we find $\delta p(\mu, \Lambda)$ to be many orders of magnitude larger than the measured value $p(\mu)$ then we suspect that we do not understand the physics at energies $\mu < E < \Lambda$ as well as we thought. This is actually the case for the cosmological constant: cosmological observations indicate that the vacuum's energy density is at present $\rho \sim (10^{-3}eV)^4$ but the theoretical prediction for $\delta \rho(\mu, \lambda)$ is many orders of magnitude greater as a particle of mass m contributes an amount of $\delta \rho(\mu, \Lambda) \sim m^4$ when it is integrated out and pratically all of the elementary particles we know have $m \gg 10^{-3}eV$.

A solution of the cosmological constant problem could involve a modification of gravity at energy scale[10] $E > \mu \sim 10^{-3}eV$, but, if this is the case, must not ruin the precise agreement with all the many non-gravitational experiments which have been performed so far. A scenario which could have both these properties is the LED scenario. Indeed LED allows that only gravity propagate in large extra dimensions, while all the non gravitational interactions should be confined on a 3-brane within the extra-dimensional space. Since we have a precise relation between the 4D Planck scale, the fundamental scale M of the higher dimensional theory and the volume of the internal space we can have a constraint on the latter by requiring M to be of the order of TeV. This constraint rules out just one large extra dimension and the simplest choice is therefore a 6D model. In this case the typical physical size r of the internal space is of the order of the observed value $\rho^{-1/4}(\mu)$ of the vacuum energy density. Therefore we expect that grav-

---

[10]Other interesting works [28, 29, 30] propose a possible solution of the cosmological constant problem, by considering modification of gravity at very large distances, as we have summarized in Section 1.5.

ity (and also its prediction for ρ) is much different from ordinary 4D Einstein's theory at submillimeter length scale, which corresponds to energies close to $10^{-3}eV$. The idea of SLED is to supersymmetrize a LED model in order to have control on the prediction for ρ: indeed we know that a supersymmetric theory, without explicit and spontaneous symmetry breaking, predicts a vanishing vacuum's energy. One can hope to get the observed value of the vacuum's energy performing a small supersymmetry breaking in such a framework.

In a higher dimensional theory with size r of extra dimensions the superymmetry breaking scale is expected to be of the order of $1/r$, and therefore small for large extra dimensions. Of course in order for this to be phenomenologically viable one has to find a mechanism which separates the supersymmetry breaking in the bulk and on the brane. The complete (theoretical) value of ρ in this context can be computed by performing the following three steps. First one integrates out at the full quantum level all the brane degrees of freedom to obtain an effective theory defined on the brane with tension T. After that one can consider such tension as source for the higher dimensional geometry by inserting in Einstein equation delta-function sources proportional to T. In a 6D supergravity the sum of the quantum brane contribution and the classical bulk contribution to ρ cancel exactly and one is left with only the bulk quantum contribution. In some circumstances this quantum contribution is of order m_{sb}^4, where $m_{sb} \sim M^2/M_{Pl}$, where M is the 6D Planck mass and M_{Pl} is the 4D Planck mass [79]. The small size of the 4D vacuum energy is in this way attributed to the very small size with which supersymmetry breaks in the bulk relative to the scale with which it breaks on the brane.

Although this is an interesting idea, there are some points that are not clear and need more investigations, for instance the complete effect of supersymmetry breaking in the bulk and of the compactness of the internal space, which is not shared by the models [28, 29, 30] that we have briefly summarized at the end of Section 1.5. Moreover, in explicit realizations of the SLED scenario, for example in the original paper [34], a background monopole is introduced on the bulk to support a compact internal manifold. Consequently a Dirac quantization condition in general emerges and it is not completely clear if this additional constraint can ruin the SLED argument. Finally, a detailed study of such a scenario requires the complete analysis of perturbations, which are actually needed in order to compute sistematically all the contributions to the vacuum's energy density.

However, in order to implement this idea, one is interested in explicit brane solutions of the 6D supergravity EOM, which break supersymmetry. Such solutions are presented in Section 4.4.

4.4 Brane Solutions

So far we have analysed smooth solutions of minimal gauged supergravity, which turn out to be also the maximal symmetric one. In this section we want to study what happens if one relaxes the maximal symmetry assumption for the complete 6D space-time. As we will see this necessarily leads to singularities. One interesting application is interpreting them as 3-branes which support SM fields. Indeed in order to realize a LED or a SLED scenario brane solutions are interesting because standard KK compactification are known to be phenomenological incompatible with LED.

However, we shall assume the following properties of the background solutions:

(i) *4D Poincaré invariance.*

(ii) *Axisymmetry of the internal 2D space.*

(iii) *The hyperscalars are not active.*

Assumptions (i), (ii) and (iii) will simplify our calculations, but (iii) is also motivated by the fact that the potential has a global minimum in $\phi^\alpha = 0$, and therefore such set up will support stability. The 4D Poincaré invariance implies that the Kalb-Ramond field strength vanishes, and this, together with (iii), leads to the bosonic EOM (4.2.31). For later purposes, we remark that those equations are invariant under the constant classical scaling symmetry

$$G_{MN} \to \xi G_{MN}, \qquad e^{\kappa\sigma/2} \to \xi e^{\kappa\sigma/2}. \qquad (4.4.36)$$

Furthermore, the sole effect of the transformation

$$e^{\kappa\sigma/2} \to \zeta^2 e^{\kappa\sigma/2}, \qquad \left(F_{1MN}, \tilde{F}_{MN}\right) \to \zeta^{-1}\left(F_{1MN}, \tilde{F}_{MN}\right) \qquad (4.4.37)$$

is the rescaling

$$g_1 \to \zeta g_1. \qquad (4.4.38)$$

We have to remember these symmetries when we will count the number of independent parameters for a given solution. Now we analyse the implications of (i) and (ii) on the background tensors. The 6D metric has to be of the form

$$ds^2 = e^{A(\rho)}\eta_{\mu\nu}dx^\mu dx^\nu + d\rho^2 + e^{B(\rho)}d\varphi^2, \qquad (4.4.39)$$

where ρ is a *radial coordinate*, with range $0 \le \rho \le \bar{\rho}$, and φ an *angular coordinate*, whose range is assumed to be $0 \le \varphi < 2\pi$, moreover A and B does not depend on φ and on the 4D coordinates x^μ. Another equivalent radial coordinate is

$$u(\rho) \equiv \int_0^\rho d\rho' e^{-A(\rho')/2}, \qquad (4.4.40)$$

whose range is $0 \le u \le \bar{u} \equiv \int_0^{\bar{\rho}} d\rho e^{-A(\rho)/2}$, and in this frame the metric reads

$$ds^2 = e^{A(u)}\left(\eta_{\mu\nu}dx^\mu dx^\nu + du^2\right) + e^{B(u)}d\varphi^2. \qquad (4.4.41)$$

The coordinate u will be useful when we will discuss fluctuations in Chapter 5. Indeed u will turn out to be the independent variable of 1D Schroedinger-like equation governing the fluctuations. The metric (4.4.39) represents a warped geometry because of the warp factor e^A. Moreover the dilaton and the gauge field depend only on ρ and

$$F_{\mu\nu} = 0, \quad F_{\mu m} = 0, \quad F_{mn} = f(\rho)\epsilon_{mn}, \qquad (4.4.42)$$

where ϵ_{mn} is the 2D anti-symmetric Levi-Civita symbol (with $\epsilon_{\rho\varphi} = 1$).

4.4.1   General Axisymmetric Solutions

In Ref. [63] Gibbons, Guven and Pope (GGP) found the most general solution satisfying (i), (ii), and (iii). We do not prove that the GGP solution is actually a solution because this would involve a very long and standard calculation but we give now the explicit expression:

$$
\begin{aligned}
e^{2A} &= \left(\frac{q\lambda_2}{4g_1\lambda_1}\right)\frac{\cosh\left[\lambda_1(\eta-\eta_1)\right]}{\cosh\left[\lambda_2(\eta-\eta_2)\right]}, \\
e^{-2B} &= \left(\frac{g_1q^3}{\lambda_1^3\lambda_2}\right)e^{-2\lambda_3\eta}\cosh^3\left[\lambda_1(\eta-\eta_1)\right]\cosh\left[\lambda_2(\eta-\eta_2)\right] \\
F &= \left(\frac{gq}{\kappa}\right)e^{B-A}e^{-\lambda_3\eta}Qd\eta\wedge d\varphi, \\
e^{\kappa\sigma} &= e^{2A}e^{2\lambda_3\eta},
\end{aligned}
\tag{4.4.43}
$$

where $\eta \equiv \int^\rho d\rho' e^{-2A-B/2}$, Q again denotes a generator of a $U(1)$ subgroup of a simple factor of $\mathcal{G}$, satisfying $Tr\left(Q^2\right) = 1$, moreover $q, \eta_{1,2}, \lambda_{1,2,3}$ are constants, and the parameters λ_i satisfy

$$
\lambda_1^2 + \lambda_3^2 = \lambda_2^2.
\tag{4.4.44}
$$

Without loss of generality one can take $\lambda_{1,2} \geq 0$ and Eq. (4.4.44) implies $\lambda_2 \geq \lambda_1$. Moreover in (4.4.43) F represents the gauge field strength which is not zero in the background and g the corresponding gauge constant. At face value we have a total of 5 integration constants $q, \lambda_1, \lambda_2, \eta_1, \eta_2$. However, one combination of these five constants correspond to the scaling (4.4.36). A second combination similarly corresponds to the second rescaling (4.4.37), whose sole effect is (4.4.38), leaving a total of 3 nontrivial parameters. The metric in (4.4.43) has at most 2 singularities because singularities can occur where the metric components vanish or diverge. Inspection of (4.4.43) shows that this only occurs when $\eta \to -\infty$ and $\eta \to +\infty$. Such singularities are not all of the purely conical type. The metric singularities which are not purely conical ($\lambda_3 \neq 0$) come in two categories [66]. Some can be interpreted as describing the fields of two localized 3-brane like objects, while others are better interpreted as the bulk fields which are sourced by a combination of a 3-brane and 4-brane, rather than being due to two 3-branes. An extended discussion on the singularities of the non purely conical type is given in [66]. In the next subsection we shall describe the purely conical case.

The class of solutions that we presented here are not the most general appearing in the literature. Relaxing the condition of 4D Poincaré symmetry to that of only 4D maximal symmetry should allow more general solutions to be found [80]. More general solutions also exist breaking axial symmetry [69], or having nontrivial VEVs for the hyperscalars [71]. Furthermore, *dyonic string* solutions have been constructed in [65].

4.4.2   Conical Singularities

Here we focus on the subset of the general solutions (4.4.43) having purely conical singularities ($\lambda_3 = 0$), which can be interpreted as being sourced by two 3-branes. We shall call this background conical-GGP solution.

The explicit conical-GGP solution is then[11] [63, 66]:

$$
e^A = e^{\kappa\sigma/2} = \sqrt{\frac{f_1}{f_0}}, \quad e^B = \alpha^2 e^A \frac{r_0^2 \cot^2(u/r_0)}{f_1^2},
$$
$$
\mathcal{A} = -\frac{4\alpha g}{q\kappa f_1} Q \, d\varphi, \tag{4.4.45}
$$

where α is a real positive number, q is a real number, and moreover

$$
f_0 \equiv 1 + \cot^2\left(\frac{u}{r_0}\right), \quad f_1 \equiv 1 + \frac{r_0^2}{r_1^2} \cot^2\left(\frac{u}{r_0}\right), \tag{4.4.46}
$$

with $r_0^2 \equiv \kappa^2/(2g_1^2)$, $r_1^2 \equiv 8/q^2$. The range of u is from 0 to $\bar{u} \equiv \pi r_0/2$ and the deficit angles are

$$
\delta = 2\pi\left(1 - \alpha\frac{r_1^2}{r_0^2}\right). \tag{4.4.47}
$$
$$
\bar{\delta} = 2\pi\left(1 - \alpha\right). \tag{4.4.48}
$$

Notice then that a non-trivial warping enforces the presence of a 3-brane source. On the other hand, the parameter α is not fixed by the EOM and it represents a modulus.

The expression for the gauge field background in equation (4.4.45) is well-defined in the limit $u \to 0$, but not as $u \to \bar{u}$. We should therefore use a different patch to describe the $u = \bar{u}$ brane, and this must be related to the patch including the $u = 0$ brane by a single-valued gauge transformation[12]. This leads to a Dirac quantization condition, which for a field interacting with $\mathcal{A}$ through a charge e

---

[11]The coordinate u is related to the coordinate r in [63] by $r = r_0 \cot(u/r_0)$.

[12]We have already treated this topological constraint for the sphere compactification of the Einstein-Maxwell-Scalar system in Subsection 3.2.1 of Chapter 3.

gives

$$\alpha e \frac{4g}{\kappa q} = \alpha e \frac{r_1}{r_0} \frac{g}{g_1} = N \qquad (4.4.49)$$

where N is an integer. The charge e can be computed once we have selected the background gauge group, since it is an eigenvalue of the generator Q. Finally, an explicit calculation shows that the internal manifold corresponding to solution (4.4.45) has an S^2 topology (its Euler number equals 2). Therefore the internal manifold is compact. Thus we expect a discrete spectrum of the fluctuations around such a background. This is confirmed by the calculation of Chapter 5, where we will focus on bulk sectors that could give rise to SM-like gauge fields and charged matter.

Chapter 5

Fluctuations around Brane Solutions

In Chapter 4 we have described minimal 6D gauged supergravity and its solutions which have received much interest recently for several reasons. From the top down, the theory shares many features in common with 10D supergravity, whilst remaining relatively simple, and so it can be used as a toy model for 10D string theory compactifications. From the bottom up, it provides a context in which to extend the well-trodden path of 5D brane world models to codimension two. Moreover, as we have discussed in Section 4.3, in 6D models, SLED have shown some promise in addressing the two fine-tuning problems of fundamental physics: the Gauge Hierarchy and the Cosmological Constant Problems.

In terms of the phenomenological study of brane worlds, one should ask what are the qualitative differences between 5D and 6D models. For example, in 5D RS models, that we have summarized in Section 1.4, the warping of 4D spacetime slices is exponentially dependent on the proper radius of the extra dimension, whereas in the six dimensional models of Chapter 4 it is only power law dependent, at least if we consider the brane solutions given in Section 4.4. Moreover, the singularities sourced by the branes are distinct, the codimension one case being a jump and the codimension two case being conical.

5D RS models with a large (or infinite) extra dimension and the SM confined to the brane were developed to explain the hierarchy in the Planck and Electroweak scales. Although the mass gap in the KK spectrum goes to zero as usual in the infinite volume limit, 4D physics is retrieved thanks to the warp factor's localization of the zero mode graviton - and exponential suppression of higher modes - close to the brane with positive tension. Subsequently it was found that the step singularities in the geometry could also localize bulk fermions [25], in much the same way as previously achieved with scalar fields and kink topological defects, that we have introduced in Subsection 1.2.1. Models were then developed in which SM fields all[1] originate from the bulk as localized degrees of freedom [26].

A study of warped brane worlds in 6D supergravity has been given in Section 4.4, where the focus was on the background solutions. A general solution with 4D Poincaré and 2D axial symmetry was given, and it was shown in Subsection 4.4.2 that warping can lead to conical singularities in the internal manifold, which can be naturally interpreted as 3-branes sources. It is certainly interesting to go beyond these background solutions, and study the dynamics of their fluctuations. The final objective would be to obtain the effective theory describing 4D physics, and an understanding of when this effective theory is valid.

Although in these constructions the SM is usually put in by hand, envisioned on a 3-brane source, the bulk theory is potentially rich enough to contain the SM gauge and matter fields. As hinted above, should the SM arise as KK zero modes of bulk fields, there are two ways to hide the heavy modes and recover 4D physics. They may have a large mass gap, and thus be unattainable at the energy scales thus-far encountered in our observed universe. Or they may be light but very weakly coupled to the massless modes, for example if the massless modes are peaked near to the brane, and the massive modes are not. In any case, whether or not one expects the bulk to give rise to the SM, one should study its degrees of freedom and determine under which conditions they are observable or out of sight. This could also prove useful for a deeper understanding of the self-tuning mechanism of SLED, and its quantum corrections.

A complete study of the linear perturbations is a very complicated problem, involving questions of gauge-fixings and a highly coupled system of dynamical equations. Some partial results have been obtained for the scalar perturbations in [81]. In this chapter we consider sectors

---

[1]Although the Higgs field should be confined to the brane in order not to lose the gauge hierarchy.

within which SM gauge and charged matter fields might be found. By some fortune, these also happen to be two of the least complicated ones. Much of our discussion is general, and could easily be applied or extended to other 6D models with axial symmetry. We follow the usual KK procedure, and reduce the equations of motion to an equivalent non-relativistic quantum mechanics problem, which we are able to solve exactly. We consider carefully the boundary conditions that the physical modes must satisfy, and from these derive the wave function profiles and complete discrete mass spectra.

Our exact solutions enable us to analyze in detail the effects of the power-law warping and conical defects that arise in 6D brane worlds. We find that the warping cannot give rise to zero modes peaked at the brane, without also leading to peaked profiles for the entire KK tower. On the other hand, the conical defects do break another standard lore of the classical KK theory. Remarkably, even if the volume of the internal manifold goes to infinity, the mass gap does not necessarily go to zero. This decoupling between the mass gap and volume means that in principle SM fields, in addition to gravity, could 'feel' the extent of large extra dimensions, whilst still being accurately described by a 4D effective field theory.

This chapter is organized as follows. We begin in Section 5.1 by giving our set up and additional properties of axisymmetric solutions, that we have introduced in Section 4.4. Then, in Section 5.2 we analyze the gauge field fluctuations, deriving the wave functions and masses of the KK spectrum. A similar analysis is presented in Section 5.3 for the fermions. Section 5.4 discusses the physical implications of the results found, and in particular whether they can be naturally applied to the LED scenario. Finally we end in Section 5.5 with some conclusions and future directions.

In the appendices we give some results that are useful for the detailed calculations. Appendix C.2 explains how the conical defects manifest themselves in the metric ansatz. In Appendix C.3 we show in detail how the boundary conditions are applied to obtain a discrete mass spectrum, and we give the complete fermionic mass spectrum thus derived in Appendix C.4.

5.1   The Set Up

We consider 6D $N = (1,0)$ gauged supergravity, and its warped braneworld solutions, whose fluctuations we will then study. This theory has been defined in Section 4.2, in particular the field content is given in (4.2.3)-(4.2.10). As we discussed in Subsection 4.2.2, in general the theory has anomalies but for certain gauge groups and hypermultiplet representations these anomalies can be cancelled via a Green-Schwarz mechanism. We will consider a general matter content, with gauge group of the form $\mathcal{G} = \tilde{\mathcal{G}} \times U(1)_R$. For example, we could take the anomaly free group $\mathcal{G} = E_6 \times E_7 \times U(1)_R$, under which the fermions are charged as follows: $\psi_M \sim (1,1)_1$, $\chi \sim (1,1)_1$, $\lambda \sim (78,1)_1 + (1,133)_1 + (1,1)_1$, $\psi \sim (1,912)_0$.

We remind that the bosonic action and the EOM take the form (4.2.30), and (4.2.31) respectively. In this chapter we will consider the general class of warped solutions with 4D Poincaré symmetry, and axial symmetry in the transverse dimensions that we described in Section 4.4. We can summarize those solutions as follows:

$$
\begin{aligned}
ds^2 = G_{MN}dX^M dX^N &= e^{A(\rho)}\eta_{\mu\nu}dx^\mu dx^\nu + d\rho^2 + e^{B(\rho)}d\varphi^2, \\
\mathcal{A} &= \mathcal{A}_\varphi(\rho)Qd\varphi, \\
\sigma &= \sigma(\rho), \\
G_{MNP} &= 0, \quad\quad\quad\quad (5.1.1)
\end{aligned}
$$

with $0 \leq \rho \leq \bar{\rho}$ and $0 \leq \varphi < 2\pi$. In the following we shall also use the radial coordinate u defined in (4.4.40), and in this frame the metric has the form given in (4.4.41).

Given the above ansatz, the general solution to the equations of motion (4.2.31) is given in Subsection 4.4.1. Although much of our formalism for the perturbation analysis can be applied to the general ansatz (5.1.1), we will focus on a subset of this general solution, namely that which contains singularities no worse than conical. Thus, in addition to the ansatz (5.1.1), we impose the following asymptotic behaviour for the metric:

$$
e^A \overset{\rho\to 0}{\to} constant \neq 0, \quad e^A \overset{\rho\to\bar{\rho}}{\to} constant \neq 0,
$$

and

$$e^{B} \overset{\rho \to 0}{\to} (1 - \delta/2\pi)^2 \, \rho^2, \quad e^{B} \overset{\rho \to \bar{\rho}}{\to} \left(1 - \bar{\delta}/2\pi\right)^2 (\bar{\rho} - \rho)^2, \quad (5.1.2)$$

that is we assume conical singularities with deficit angle δ at $\rho = 0$ and $\bar{\delta}$ at $\rho = \bar{\rho}$, at which points the Ricci scalar contains delta-functions (see Appendix C.2). These singularities can be interpreted as 3-brane sources with tensions $T = 2\delta/\kappa^2$ and $\bar{T} = 2\bar{\delta}/\kappa^2$ as we proved in Section 1.6. The explicit expression for these solutions is given in Subsection 4.4.2 and we called them conical-GGP solutions.

We end this section by considering the various parameters in the model, and the phenomenological constraints which can arise when we give it a brane world interpretation. There are three free parameters in the 6D theory, which can be taken to be the gauge coupling $\tilde{g}$, and two out of the following three parameters: the 6D Planck scale, κ, the gauge coupling g_1 and the length-scale $r_0 = \kappa/\sqrt{2}g_1$. In the solution there are two free parameters, r_1 (or q) and α. However, one combination of all these parameters is constrained by the quantization condition (4.4.49).

The relation between the 6D Planck scale κ and our observed 4D Planck scale κ_4 is

$$\frac{1}{\kappa^2} V_2 = \frac{1}{\kappa_4^2}, \quad (5.1.3)$$

where the volume V_2 is given by

$$V_2 = \int d^2 y \sqrt{-G} e^{-A} = 2\pi \int du \, e^{(3A+B)/2}. \quad (5.1.4)$$

This is a particular case of (1.3.40), and (1.3.41). For solution (4.4.45) we have

$$V_2 = 4\pi\alpha \left(\frac{r_0}{2}\right)^2. \quad (5.1.5)$$

Notice that this volume does not depend on r_1, and so we can keep it fixed whilst varying the warp factor, namely e^A in (4.4.45). Moreover from (5.1.3) a phenomenological constraint follows between the bulk couplings and the brane tensions, which can be written:

$$\frac{g_1}{\sqrt{\alpha}} = \sqrt{\frac{\pi}{2}} \kappa_4. \quad (5.1.6)$$

This implies that $g_1/\sqrt{\alpha}$ is very small, of the order of the Planck length.

Now let us embed the ADD scenario into the present model, in order to try to explain the large hierarchy between the Electroweak scale and the Planck scale via the size of the extra dimensions. Thus identifying the 6D fundamental scale with the Electroweak scale $\kappa \sim TeV^{-2}, (10TeV)^{-2}$ and constraining the observed 4D Planck scale $\kappa_4^2 \sim 10^{-32}TeV^{-2}$; the above relation translates to[2]:

$$\sqrt{\alpha}r_0 \sim 0.1mm. \tag{5.1.7}$$

Here, the LED corresponds to tuning the bulk gauge coupling and brane tensions. However, we can also observe that (5.1.7) fixes just one parameter among α, r_0 and r_1 and we still have two independent parameters even if we require large extra dimensions. Later we will see that this novel feature proves to have interesting consequences for the mass spectrum of fluctuations.

5.2   Gauge Fields

Having established the brane world solution and its properties, we are now ready to examine the fluctuations about this background, which will represent the physical fields in our model. In this section our focus will be on the gauge field fluctuations.

Normalizable gauge field zero modes in axially symmetric codimension two branes are known to exist [82, 83, 52]. However, in these known examples there is no mass gap between the zero and non-zero modes which renders an effective 4D description somewhat problematic, especially in non-Abelian case [84]. In contrast to this for the axisymmetric solutions studied in this chapter the presence of a mass gap will be automatic due to the compactness of the transverse space. In this section we shall give the full spectrum of zero and non-zero modes.

As we discussed in Subsection 4.2.2 in a more general context, given the symmetries of the problem, we can expect that the gauge fields in the low energy effective theory belong to $\mathcal{H} \times U(1)_{KK}$, where $\mathcal{H}$ is the unbroken subgroup of $\mathcal{G}$ that commutes with the $U(1)_M \subset \mathcal{G}$ in which the monopole lies[3]. The $U(1)_{KK}$ arises from the vector fluctuations

---

[2]We remind that the following conversion relation holds: $(TeV)^{-1} \sim 10^{-16}mm$.

[3]The gauge group $\mathcal{H}$ does not contain $U(1)_M$ because, as we discussed in Sub-

of the metric, due to the axial symmetry of the internal manifold, and is promoted to $SU(2)_{KK}$ in the sphere limit of the background.

The non-Abelian sector of $\mathcal{H}$ may be rich enough to contain the SM gauge group. For example, consider the anomaly free model of [59], with gauge group $\mathcal{G} = E_6 \times E_7 \times U(1)_R$, and the monopole background in E_6. As we pointed out in Subsection 4.2.2, the surviving gauge group, $SO(10) \times E_7 \times U(1)_R \times U(1)_{KK}$, then contains the Grand Unified Group $SO(10)$, and the model also includes charged matter in the fundamental of $SO(10)$. Therefore our present interest will be in the fields belonging to various representations of $\mathcal{H}$. Specifically, we will consider gauge field fluctuations orthogonal to the monopole background. For the case $\mathcal{G} = E_6 \times E_7 \times U(1)_R$, the gauge field sectors that are covered by our analysis are given in Table 5.1.

5.2.1 Kaluza-Klein Modes

Using the background solution in the 6D action (4.2.30), we can identify the bilinear action for the fluctuations. This step requires some care, because to study the physical spectrum we must first remove the gauge freedoms in the action due to 6D diffeomorphisms and gauge transformations. The problem has been studied in a general context in [52], where the authors choose a light-cone gauge fixing.

In the light-cone gauge, the action for the gauge field fluctuations, orthogonal to the monopole background, at the bilinear level reads [52]

$$
S_G(V, V) \equiv - \int d^6 X \sqrt{-G}\, \frac{1}{2} e^\phi \big(\partial_\mu V_j \partial^\mu V^j + e^{-A} \partial_\rho V_j \partial_\rho V_j
$$
$$
+ \nabla_\varphi V_j \nabla^\varphi V^j \big), \tag{5.2.1}
$$

where V_j is the gauge field fluctuation in the light cone gauge (j=1,2) and all the indices in (5.2.1) are raised and lowered with the ρ-dependent metric G_{MN} given in (5.1.1). Indeed, here and below G_{MN} represents the background metric. We have multiplied the formula of [52] by an overall e^ϕ, with $\phi \equiv \kappa\sigma/2$ and σ in the background, due to the presence of the dilaton in our theory[4]. Notice that since we are looking at

---

section 4.2.2, the latter is broken in the 4D effective theory.

[4]That the dilaton invokes only this simple change with respect to Ref. [52] can be seen as follows. First, notice that since we are considering fluctuations orthogonal to $U(1)_M$ background, there are no mixings with other sectors, and

the sector orthogonal to the monopole background, the Chern-Simons term does not contribute, and the action takes a simple form.

In general, the covariant derivative $\nabla_\varphi V_j$ includes the gauge field background

$$\nabla_\varphi V_j = \partial_\varphi V_j + i e_V \mathcal{A}_\varphi V_j, \qquad (5.2.2)$$

where again the charge e_V can be computed using group theory once the gauge group $\tilde{\mathcal{G}}$ is chosen. The value $e_V = 0$ corresponds to the gauge fields in the 4D low energy effective theory. However, since we can do so without much expense, we keep a generic value of e_V. Those fluctuations with $e_V \neq 0$ corresponds to vector fields in a non-trivial representation of the 4D effective theory gauge group. The Dirac quantization condition (4.4.49) then gives $e_V\, 4\alpha g/(\kappa q) = N_V$, where N_V is an integer.

Next we perform a KK expansion of the 6D fields. Since our internal space is topologically S^2, we require gauge fields to be periodic functions of φ:

$$V_j(X) = \sum_m V_{jm}(x) f_m(\rho) e^{im\varphi}, \qquad (5.2.3)$$

where m is an integer.

If we put (5.2.3) in (5.2.1) we obtain kinetic terms for the 4D effective fields proportional to

$$\int d^4x \sum_m \eta^{\mu\nu} \partial_\mu V_{jm}^\dagger \partial_\nu V_{jm} \int d\rho\, e^{\phi + B/2} |f_m|^2. \qquad (5.2.4)$$

Therefore physical fluctuations, having a finite kinetic energy, must satisfy the following normalizability condition (NC):

$$\int du |\psi|^2 < \infty, \qquad (5.2.5)$$

---

the bilinear action is simply $S_G = -1/4 \int d^6 X \sqrt{-\bar{G}} e^\phi G^{MN} G^{PQ} \left(F_{MP} F_{NQ}\right)^{(2)}$. We emphasise that G_{MN} and ϕ now signify the background fields. Also, $()^{(2)}$ indicates the bilinear part in the fluctuations. Next, make the change of coordinates, $d\rho = e^{-\phi/2} d\tilde{\rho}$, and rewrite the background metric in (5.1.1) as $ds^2 = e^{-\phi} \left(e^{\tilde{A}} \eta_{\mu\nu} dx^\mu dx^\nu + d\tilde{\rho}^2 + e^{\tilde{B}} d\varphi^2 \right)$, with $\tilde{A} \equiv A + \phi$ and $\tilde{B} \equiv B + \phi$. In this way, the bilinear action, S_G, reduces to exactly the same form as that of Ref. [52], and we can proceed as they do to transform into light-cone coordinates, fix the light-cone gauge, and eliminate redundant degrees of freedom using their equations of motion.

where

$$\psi = e^{(2\phi+A+B)/4} f_m. \tag{5.2.6}$$

The quantity $|\psi|^2$ represents the probability density of finding a gauge field in $[u, u + du]$.

In fact, this is not the only condition that the physical fields must satisfy. If we want to derive the EOM from (5.2.1) through an action principle we have to impose[5] the following boundary condition

$$\int d^6 X \partial_M \left(\sqrt{-G} e^{\phi-A} V_j \mathcal{D}^M V_j \right) = 0, \tag{5.2.7}$$

where $\mathcal{D}_M$ is the gauge covariant derivative. Equation (5.2.7) represents conservation of current $J_M = e^{\phi-A} V_j \mathcal{D}_M V_j$ and it is the generalization of Eq. (1.2.16) that we studied in the simple domain wall model. Moreover, since the fields are periodic functions of φ, (5.2.7) becomes

$$\left[\sqrt{-G}\, e^{\phi-A}\, V_j \partial_\rho V_j \right]_0^{\bar{\rho}} = 0. \tag{5.2.8}$$

The EOM can then be derived as:

$$\sqrt{-G} e^{\phi-2A} \eta^{\mu\nu} \partial_\mu \partial_\nu V_j = -\partial_\rho \left(\sqrt{-G} e^{\phi-A} \partial_\rho V_j \right) - \sqrt{-G} e^{\phi-A-B} \nabla_\varphi^2 V_j. \tag{5.2.9}$$

By inserting (5.2.3) in (5.2.9) we obtain

$$-\frac{e^{-\phi+2A}}{\sqrt{-G}} \partial_\rho \left(\sqrt{-G} e^{\phi-A} \partial_\rho f_m \right) + e^{A-B} \left(m + e_V \mathcal{A}_\varphi \right)^2 f_m = M_{V,m}^2 f_m, \tag{5.2.10}$$

where $M_{V,m}^2$ are the eigenvalues of $\eta^{\mu\nu} \partial_\mu \partial_\nu$.

At this stage, we can already identify the massless fluctuation that is expected from symmetry arguments. For $e_V = 0$, when $m = 0$, a constant f_0 is a solution of (5.2.10) with $M_{V,0}^2 = 0$. This solution corresponds to 4D effective theory gauge fields. It has a finite kinetic energy, and trivially satisfies (5.2.8). The fact that such gauge fields have a constant transverse profile guarantees charge universality of fermions in the 4D effective theory (see below and Subsection 1.2.3).

To find the massive mode solutions, we can express (5.2.10) in

---

[5] Actually we impose that for every pair of fields V_j and V_j' the condition $\int d^6 X \partial_M \left(\sqrt{-G} e^{\phi-A} V_j \mathcal{D}^M V_j' \right) = 0$ is satisfied but in (5.2.7) the prime is understood.

terms of u and ψ and obtain a Schroedinger equation:

$$\left(-\partial_u^2 + V\right)\psi = M_V^2\psi, \qquad (5.2.11)$$

where the "potential" is

$$V(u) = e^{A-B}\left(m + e_V\mathcal{A}_\varphi\right)^2 + e^{-(2\phi+A+B)/4}\partial_u^2 e^{(2\phi+A+B)/4}. \quad (5.2.12)$$

We want to find the complete set of solutions to (5.2.11) satisfying the NC (5.2.5) and the boundary conditions (5.2.8), which can be written in terms of u and ψ as follows

$$\left(\lim_{u\to\bar{u}} - \lim_{u\to0}\right)\left\{\psi^*\left[-\partial_u + \frac{1}{4}\left(2\partial_u\phi + \partial_uA + \partial_uB\right)\right]\psi\right\} = 0. \quad (5.2.13)$$

In order for (5.2.13) to be satisfied, both the limits $u \to 0$ and $u \to \bar{u}$ must be finite. Condition (5.2.13) ensures that the Hamiltonian in the Schroedinger equation (5.2.11) is hermitian, and so has real eigenvalues and an orthonormal set of eigenfunctions. Therefore, we shall call it the hermiticity condition (HC).

So far our analysis has been valid for all axially symmetric solutions of the form (5.1.1). We will now use these results to determine the fluctuation spectrum about the conical-GGP solution (4.4.45). We observe that $V(u)$ then contains a delta-function contribution, arising from the second-order derivative of the conical metric function $\partial_u^2 B$ (see Appendix C.2 and Eq. (C.2.6)). However, we can drop it because $\partial_u^2 B$ also contains stronger singularities at $u = 0$ and $u = \bar{u}$: respectively $1/u^2$ and $1/(\bar{u} - u)^2$. These singularities are also a consequence of the behavior of e^B given in (5.1.2) and they imply that the behaviour of the wave functions close to $u = 0$ and $u = \bar{u}$ cannot depend on the mass. In particular, this immediately implies that if the wave functions of zero modes are peaked near to one of the branes, then the same will be true also for the infinite tower of non-zero modes. In other words, we cannot hope to dynamically generate a brane world scenario, in which zero modes are peaked on the brane, and massive modes are not, leading to weak coupling between the two sectors[6]. If we are to interpret the zero mode gauge fields as those of the SM, therefore, for the massive modes to have escaped detection they must

---

[6]In fact, a similar singular behaviour for the potential in general arises for the general axisymmetric solutions given in Subsection 4.4.1 and studied in [66], where the hypothesis (5.1.2) is relaxed.

have a large mass gap.

Meanwhile, we note that in contrast to the non-relativistic quantum mechanics problem, here we cannot deduce qualitative results about the mass spectrum from the shape of the potential. This is because the boundary conditions to be applied in the context of dimensional reduction are in general different to those in problems of quantum mechanics. In particular, the HC (5.2.13) is a non-linear condition, contrary to the less general linear boundary conditions usually encountered in quantum mechanics to ensure hermiticity of the Hamiltonian. We will be able to impose the more general case thanks to the universal asymptotic behaviour of the KK tower.

Returning then to our explicit calculation of the KK spectrum, we can write $V(u)$ as

$$V(u) = V_0 + v \cot^2\left(\frac{u}{r_0}\right) + \bar{v} \tan^2\left(\frac{u}{r_0}\right), \qquad (5.2.14)$$

and

$$r_0^2 V_0 \equiv 2m\omega(m - N_V)\bar{\omega} - \frac{3}{2}, \quad r_0^2 v \equiv m^2\omega^2 - \frac{1}{4},$$
$$r_0^2 \bar{v} \equiv (m - N_V)^2 \bar{\omega}^2 - \frac{1}{4}. \qquad (5.2.15)$$

Moreover in this case the expression (5.2.13) for the HC becomes

$$\lim_{u \to \bar{u}} \psi^*\left(-\partial_u + \frac{1}{2}\frac{1}{u - \bar{u}}\right)\psi - \lim_{u \to 0} \psi^*\left(-\partial_u + \frac{1}{2u}\right)\psi = 0. \quad (5.2.16)$$

If we introduce z and y in the following way [85]

$$z = \cos^2\left(\frac{u}{r_0}\right), \qquad \psi = z^\gamma (1 - z)^\beta y(z), \qquad (5.2.17)$$

Eq. (5.2.11) becomes

$$z(1 - z)\partial_z^2 y + [c - (a + b + 1)z]\partial_z y - aby = 0, \qquad (5.2.18)$$

where

$$
\gamma \equiv \frac{1}{4}\left[1 + 2(m - N_V)\varpi\right], \quad \beta \equiv \frac{1}{4}\left(1 + 2m\omega\right), \quad c \equiv 1 + (m - N_V)\varpi,
$$

$$
a \equiv \frac{1}{2} + \frac{m}{2}\omega + \frac{1}{2}(m - N_V)\varpi
$$
$$
+ \frac{1}{2}\sqrt{r_0^2 M_{V,m}^2 + 1 + [m\omega - (m - N_V)\varpi]^2},
$$

$$
b \equiv \frac{1}{2} + \frac{m}{2}\omega + \frac{1}{2}(m - N_V)\varpi
$$
$$
- \frac{1}{2}\sqrt{r_0^2 M_{V,m}^2 + 1 + [m\omega - (m - N_V)\varpi]^2}, \tag{5.2.19}
$$

and

$$
\omega \equiv (1 - \delta/2\pi)^{-1}, \qquad \varpi \equiv (1 - \bar\delta/2\pi)^{-1}. \tag{5.2.20}
$$

Eq. (5.2.18) is the hypergeometric equation and its solutions are known. For $c \neq 1$ the general solution is a linear combination of the following functions:

$$
y_1(z) \equiv F(a, b, c, z), \quad y_2(z) \equiv z^{1-c}F(a + 1 - c, b + 1 - c, 2 - c, z), \tag{5.2.21}
$$

where F is Gauss's hypergeometric function. So for $c \neq 1$ the general integral of the Schroedinger equation is

$$
\psi = K_1\psi_1 + K_2\psi_2, \tag{5.2.22}
$$

where

$$
\psi_i \equiv z^\gamma(1 - z)^\beta y_i. \tag{5.2.23}
$$

and $K_{1,2}$ are integration constants. For $c = 1$ we have $\psi_1 = \psi_2$ but we can construct a linearly independent solution using the Wronskian method and the general solution reads

$$
\psi = K_1\psi_1 + K_2\psi_1 \int^u \frac{du'}{\psi_1^2(u')}. \tag{5.2.24}
$$

Now we must impose the NC (5.2.5) and HC (5.2.16), to select the physical modes. In Appendix C.3 we give explicit calculations; the final result is that the NC and HC give the following discrete spectrum. The wave functions are

$$\psi \propto z^\gamma (1-z)^\beta F(a,b,c,z), \quad for \quad m \geq N_V, \tag{5.2.25}$$
$$\psi \propto z^{\gamma+1-c}(1-z)^\beta F(a+1-c,b+1-c,2-c,z), \tag{5.2.26}$$
$$for \ m < N_V.$$

and the squared masses are as follows:

- For $N_V \leq m < 0$

$$M_{V\,n,m}^2 = \frac{4}{r_0^2}\left\{ n(n+1) + \left(\frac{1}{2}+n\right)[-m\omega + (m-N_V)\overline{\omega}]\right\} > 0. \tag{5.2.27}$$

- For $m \geq N_V$ and $m \geq 0$

$$M_{V\,n,m}^2 = \frac{4}{r_0^2}\left\{ n(n+1) + \left(\frac{1}{2}+n\right)[m\omega + (m-N_V)\overline{\omega}]\right.$$
$$\left. + m\omega(m-N_V)\overline{\omega}\right\} \geq 0. \tag{5.2.28}$$

- For $m < N_V$ and $m < 0$

$$M_{V\,n,m}^2 = \frac{4}{r_0^2}\left\{ n(n+1) + \left(\frac{1}{2}+n\right)[-m\omega + (N_V-m)\overline{\omega}]\right.$$
$$\left. - m\omega(N_V-m)\overline{\omega}\right\} > 0. \tag{5.2.29}$$

- For $0 \leq m < N_V$

$$M_{V\,n,m}^2 = \frac{4}{r_0^2}\left\{ n(n+1) + \left(\frac{1}{2}+n\right)[m\omega + (N_V-m)\overline{\omega}]\right\} > 0. \tag{5.2.30}$$

The masses given in (5.2.27) and (5.2.28) correspond to the wave function (5.2.25) whereas the masses given in (5.2.29) and (5.2.30) correspond to the wave function (5.2.26). We observe that there are no tachyons and that the only zero mode is for $n = 0$, $m = 0$ and $N_V = 0$ ($e_V = 0$), corresponding to gauge fields in the 4D low energy effective theory.

As a check, we can consider the S^2 limit ($\omega, \overline{\omega} \to 1$), whose mass spectrum is well-known. Our spectrum (5.2.27)-(5.2.30) reduces to

$$a^2 M_V^2 = l(l+1) - \left(\frac{N_V}{2}\right)^2, \quad multiplicity = 2l + 1, \quad (5.2.31)$$

where $a = r_0/2$ is the radius of S^2 and[7] $l = |\frac{N_V}{2}| + k$, $k = 0, 1, 2, 3, ...$
This is exactly the result that one finds by using the spherical harmonic
expansion [7] from the beginning.

At this stage we can point towards a novel property of the final
mass spectrum. Observe that in the large α (small $\bar{\omega}$) limit the volume
V_2 given in Eq. (5.1.5) becomes large but the mass gap between two
consecutive KK states does not reduce to zero as in standard KK
theories[8]. This a consequence of the shape of our background manifold
and in particular of the conical defects. Notice that the large α limit
corresponds to a negative tension brane at $u = \bar{u}$, but not necessarily
at $u = 0$.

In Section 5.3 we will show that the same effect appears also in the
fermionic sector, and we will turn to a discussion of its implications in
Section 5.4.

5.2.2   4D Effective Gauge Coupling

Let us end the discussion on gauge fields by briefly presenting the
4D effective gauge coupling. This can be obtained by dimensionally
reducing the 6D gauge kinetic term. We consider the zero mode fluctuations in $\mathcal{H}$, about the background (5.1.1), so that the 4D effective
gauge kinetic term is:

$$\int d^6 X \sqrt{-G} \left\{ -\frac{1}{4g^2} e^{\kappa\sigma/2} Tr F_{MN} F^{MN} \right\} \rightarrow$$

$$\int d^4 x \left\{ -\frac{1}{4g^2} \left[\int du d\varphi e^{(3A+B)/2} f_0^2 \right] Tr F_{\mu\nu} F^{\mu\nu} \right\} \quad (5.2.32)$$

Recalling that $f_0 = const$ and normalizing it to one, we can read:

$$\frac{1}{g_{eff}^2} = \frac{1}{g^2} V_2. \quad (5.2.33)$$

---

[7]The number l is defined in different ways in equations (5.2.27)-(5.2.30). For
instance we have $l \equiv n + |N_V/2|$ for (5.2.27).

[8]This is also true for the proper volume of the 2D internal manifold.

5.3 Fermions

We will now consider fermionic perturbations, and in particular our interest will be in the sector charged under the 4D effective gauge group, $\mathcal{H}$, discussed above. These fields arise from the hyperinos and the $\mathcal{H}$ gauginos, for which we also restrict ourselves to those orthogonal to the $U(1)_R$. Thus we are considering matter charged under the non-Abelian gauge symmetries of the 4D effective theory. For instance, for the anomaly free model $E_6 \times E_7 \times U(1)_R$, with the monopole embedded in the E_6, the gauginos in the 78 of E_6 contain a $16 + \overline{16}$ fundamental representation of the grand unified gauge group $SO(10)$, and our analysis will be applicable to them. In Tabel 5.1, we give the complete list of fermion fields that are included in our study, for the said example.

We proceed in much the same way as for the gauge field sector of the previous section, transforming the dynamical equations and necessary boundary conditions into a Schroedinger-like problem, to obtain the physical modes and discrete mass spectrum.

The bilinear action for the fluctuations of interest takes a particularly simple form, comprising as it does of the standard Dirac action:

$$S_F = \int d^6 X \sqrt{-G}\, \bar{\lambda} \Gamma^M \nabla_M \lambda, \qquad (5.3.1)$$

where[9]

$$\nabla_M \lambda = \left(\partial_M + \frac{1}{8} \omega_M^{[A,B]} [\Gamma_A, \Gamma_B] + ie\mathcal{A}_M \right) \lambda. \qquad (5.3.2)$$

Here e is the charge of λ under $U(1)_M$, and G_{MN}, $\omega_M^{[A,B]}$ and $\mathcal{A}_M$ are the background metric, spin connection and gauge field corresponding to an axisymmetric solution (5.1.1). Analogously to the gauge field analysis, in order to derive the Dirac equation

$$\Gamma^M \nabla_M \lambda = 0 \qquad (5.3.3)$$

from (5.3.1) by using an action principle, we require conservation of

---

[9]Our conventions for Γ^A and $\omega_M^{[A,B]}$ are given in Appendix A.

fermionic current:

$$\int d^6 X \partial_M \left(\sqrt{-G} \, \overline{\lambda} \Gamma^M \lambda \right) = 0. \qquad (5.3.4)$$

Eq. (5.3.4) implies that the Dirac operator $\Gamma^M \nabla_M$ is hermitian, and we shall again refer to it as the HC. This constraint is analogous to Eq. (1.2.26), which concerns fermion fluctuation around the simple kink background. Our aim is to find the complete fermionic spectrum, that is a complete set of normalizable solutions of (5.3.3) satisfying (5.3.4).

Some care is now needed when discussing the background felt by the fermionic sector in (5.3.2). As already mentioned, in order to have a correctly defined gauge connection, it is necessary to use two patches related by a single-valued gauge transformation. The same is true for the spin connection, which must be defined in such a way as to imply the conical defects in the geometry. Henceforth we focus on the patch including the $\rho = 0$ brane, chosen to be $0 \le \rho < \bar{\rho}$. For this patch a good choice for the vielbein is

$$e^a_\mu = e^{A/2} \delta^a_\mu, \quad \{e^\alpha_m\} = \begin{pmatrix} \cos \varphi & -e^{B/2} \sin \varphi \\ \sin \varphi & e^{B/2} \cos \varphi \end{pmatrix}, \qquad (5.3.5)$$

where, like in Chapter 3, a is a 4D flat index, $\alpha = 5, 6$ a 2D flat index and $m = \rho, \varphi$. The corresponding spin connection is

$$\omega^{[a,5]}_\mu = \frac{1}{2} A' e^{A/2} \delta^a_\mu \cos \varphi, \quad \omega^{[a,6]}_\mu = \frac{1}{2} A' e^{A/2} \delta^a_\mu \sin \varphi,$$

$$\omega^{[5,6]}_\rho = 0, \quad \Omega \equiv \omega^{[5,6]}_\varphi = \left(1 - \frac{1}{2} B' e^{B/2} \right), \qquad (5.3.6)$$

where $' \equiv \partial_\rho$. It can be checked that this gauge choice correctly reproduces Stokes' theorem for a small domain including the conical defect[10].

We are now ready to study the Dirac equation (5.3.3) for 6D fluctuations, and write it in terms of 4D effective fields. Since λ is a 6D Weyl spinor we can represent it by

$$\lambda = \begin{pmatrix} \lambda_4 \\ 0 \end{pmatrix}, \qquad (5.3.7)$$

---

[10]See Appendix C.2 for some steps in this calculation.

where λ_4 is a 4D Dirac spinor: $\lambda_4 = \lambda_R + \lambda_L$, $\gamma^5 \lambda_R = \lambda_R$, $\gamma^5 \lambda_L = -\lambda_L$. By using the ansatz (5.1.1), the vielbein (5.3.5), the spin connection (5.3.6) and our conventions for Γ^A in Appendix A, the Dirac equation (5.3.3) becomes

$$e^{-A/2} \gamma^\mu \partial_\mu \lambda_L = e^{i\varphi} \left[-\partial_\rho - ie^{-B/2} \left(\partial_\varphi + ie\mathcal{A}_\varphi \right) \right.$$
$$\left. - A' + \frac{1}{2}\Omega e^{-B/2} \right] \lambda_R, \tag{5.3.8}$$

$$e^{-A/2} \gamma^\mu \partial_\mu \lambda_R = e^{-i\varphi} \left[\partial_\rho - ie^{-B/2} \left(\partial_\varphi + ie\mathcal{A}_\varphi \right) \right.$$
$$\left. + A' - \frac{1}{2}\Omega e^{-B/2} \right] \lambda_L. \tag{5.3.9}$$

Performing the Fourier mode decomposition:

$$\lambda_4(X) = \lambda_R(X) + \lambda_L(X) = \sum_m \left(\lambda_{R,m}(x) f_{R,m}(\rho) \right.$$
$$\left. + \lambda_{L,m}(x) f_{L,m}(\rho) \right) e^{im\varphi}, \tag{5.3.10}$$

where m is an integer, and inserting into (5.3.8) and (5.3.9) we find:

$$e^{-A/2} \gamma^\mu \partial_\mu \lambda_{L,m+1} f_{L,m+1} = \left[-\partial_\rho + e^{-B/2} \left(m + \frac{1}{2}\Omega + e\mathcal{A}_\varphi \right) - A' \right]$$
$$\times \lambda_{R,m} f_{R,m}, \tag{5.3.11}$$

$$e^{-A/2} \gamma^\mu \partial_\mu \lambda_{R,m-1} f_{R,m-1} = \left[\partial_\rho + e^{-B/2} \left(m - \frac{1}{2}\Omega + e\mathcal{A}_\varphi \right) + A' \right]$$
$$\times \lambda_{L,m} f_{L,m}. \tag{5.3.12}$$

For the boundary conditions, analogously to the gauge fields, the NC can be found to be:

$$\int du \, |\psi|^2 < \infty \tag{5.3.13}$$

where

$$\psi \equiv e^{A+B/4} f_{Rm} \tag{5.3.14}$$

and a similar condition for left-handed spinors. Meanwhile, the HC

(5.3.4) can be written:

$$\left[\sqrt{-G}\, f_{L,m+1} f_{R,m}^*\right]_0^{\bar{\rho}} = 0. \tag{5.3.15}$$

Having set up the dynamical equations and the relevant boundary conditions, we shall now use this information to study the complete fermionic spectrum in Subsections 5.3.1 and 5.3.2. In particular, we will focus on the questions of wave function localization, and the mass gap problem, crucial to the development of a phenomenological brane world model.

5.3.1 Zero Modes

We begin by finding the zero mode solutions, for which the problem simplifies considerably. Indeed, for the zero modes $\gamma^\mu \partial_\mu = 0$, and the equations for right- and left-handed modes (5.3.11) and (5.3.12) decouple:

$$\left[\partial_\rho - e^{-B/2}(m + e\mathcal{A}_\varphi) + A' - \frac{1}{2}\Omega e^{-B/2}\right] f_{R,m} = 0, \tag{5.3.16}$$

$$\left[\partial_\rho + e^{-B/2}(m + e\mathcal{A}_\varphi) + A' - \frac{1}{2}\Omega e^{-B/2}\right] f_{L,m} = 0. \tag{5.3.17}$$

By using the expression for Ω in equation (5.3.6), the solution of (5.3.16) is

$$f_{R,m}(\rho) \propto \exp\left[-A - \frac{1}{4}B + \int^\rho d\rho' e^{-B/2}\left(m + \frac{1}{2} + e\mathcal{A}_\varphi\right)\right], \tag{5.3.18}$$

whereas the solution of (5.3.17) can be obtained by replacing $m, e \to -m, -e$ in (5.3.18). The solution (5.3.18) for $e = 0$ was found in [86]. Here we give the expression for every e because we want to include charged fermions. We note that the zero mode solution (5.3.18) automatically satisfies the HC given in (5.3.15). From (5.3.18), (5.3.14) and (4.4.40) we obtain

$$\psi \propto \exp\left[\int^u du' e^{(A-B)/2}\left(m + \frac{1}{2} + e\mathcal{A}_\varphi\right)\right]. \tag{5.3.19}$$

For the conical-GGP background, (4.4.45), the explicit expression for ψ is

$$\psi \propto \sin\left(\frac{u}{r_0}\right)^{\omega(1/2+m)} \cos\left(\frac{u}{r_0}\right)^{\overline{\omega}(N-1/2-m)}, \tag{5.3.20}$$

where we used (4.4.49). The NC is satisfied when

$$\frac{\delta}{4\pi} - 1 < m < N - \frac{\overline{\delta}}{4\pi}. \tag{5.3.21}$$

From here we retrieve the result that for the sphere, which has $\delta = \overline{\delta} = 0$, there exist normalizable zero modes only for $e \neq 0$ ($N \neq 0$), that is for a non-vanishing monopole background [7]. Moreover, as found in [86], we see that the conical defects also make massless modes possible, provided that there is at least one negative deficit angle, even if $N = 0$. However, (5.3.21) implies that for positive tension branes, $\delta, \overline{\delta} > 0$, the adjoint of $\mathcal{H}$, which has $e = 0$, is projected out. If $\mathcal{H}$ contains the SM gauge group, this is appealing since the fermions of the SM are not in adjoint representations. In any case, the number of families depends on δ, $\overline{\delta}$ and N.

Let us now consider the wave function profiles, (5.3.20). Observe that ψ is peaked on the $u = 0$ brane (that is, $\psi \to \infty$ as $u \to 0$, and $\psi \to 0$ as $u \to \bar{u}$) when

$$m < -1/2, \quad and \quad m < -1/2 + N. \tag{5.3.22}$$

By comparing (5.3.21) and (5.3.22) we understand that we have normalizable and peaked ψ only for $\delta < 0$ (that is for negative tension brane). If $N - \overline{\delta}/4\pi > 0$ we can have normalizable zero modes for $\delta > 0$ (positive tension brane) but the corresponding ψ are not peaked on the $u = 0$ brane (indeed, $\psi \to 0$ as $u \to 0$). On the other hand from (5.3.20) we see that ψ is peaked on the $u = \bar{u}$ brane when

$$m > -1/2, \quad and \quad m > -1/2 + N. \tag{5.3.23}$$

By comparing (5.3.21) and (5.3.23) we understand that we have normalizable and peaked ψ only for $\overline{\delta} < 0$ (that is for negative tension brane).

We can also analyze the chirality structure. Since the left handed wave functions can be obtained from (5.3.20) by replacing $m, N \to$

$-m, -N$, in order for them to be peaked on the $u = 0$ brane we need

$$m > 1/2, \quad and \quad m > 1/2 + N, \qquad (5.3.24)$$

whereas in order for the left handed wave functions to be peaked on the $u = \bar{u}$ brane we need

$$m < 1/2, \quad and \quad m < 1/2 + N. \qquad (5.3.25)$$

So, if we were to require that ψ be peaked on a brane, we always have a chiral massless spectrum because the chirality index counts the difference of modes in $f_{R,m}$ and $f_{L,m}$ with given m, $N_R(m) - N_L(m)$ [86]. We should point out, however, that in fact peaked zero modes, ψ, may not be necessary in order to have an acceptable phenomenology. The answer to this question can be found only after constructing the complete spectrum, and studying the couplings between different 4D effective fields.

5.3.2   Massive Modes

We now move on to a study of the complete KK tower for the fermions. We begin by establishing the corresponding Schroedinger problem. The two coupled first order ODEs, equations (5.3.11) and (5.3.12), can be equivalently expressed as a single second order ODE and a constraint equation, as follows[11]

$$e^A \left(-\partial_\rho^2 + h \partial_\rho + g_m \right) f_{R,m} = M_{F,m}^2 f_{R,m}, \qquad (5.3.26)$$

$$M_{F,m} f_{L,m+1} = e^{A/2} \left[-\partial_\rho - A' + \left(m + \frac{1}{2} \Omega \right. \right.$$
$$\left. \left. + e \mathcal{A}_\varphi \right) e^{-B/2} \right] f_{R,m}, \qquad (5.3.27)$$

---

[11] Eq.(5.3.26) can also be obtained by squaring the 6D Dirac operator, and using that $[\nabla_M, \nabla_N] = \frac{1}{4} R_{MN}{}^{AB} \Gamma_{AB} + ie F_{MN}$.

where $M_{F,m}^2$ are the eigenvalues of $(\gamma^\mu \partial_\mu)^2$ and

$$h \equiv -\frac{5}{2}A' + (\Omega - 1)e^{-B/2}, \tag{5.3.28}$$

$$
\begin{aligned}
g_m \equiv & \left[\frac{1}{2}\Omega' - \frac{1}{4}\Omega B' - \frac{m}{2}B' + \frac{5}{4}A'\Omega + \left(\frac{m}{2} - 1\right)A' \right. \\
& \left. - \frac{e}{2}B'\mathcal{A}_\varphi + e\mathcal{A}'_\varphi + \frac{e}{2}A'\mathcal{A}_\varphi \right] e^{-B/2} \\
& + \left[m(m+1) + \frac{1}{2}\Omega - \frac{\Omega^2}{4} + (2m+1)e\mathcal{A}_\varphi + e^2\mathcal{A}_\varphi^2 \right] e^{-B} \\
& - A'' - \frac{3}{2}(A')^2. \tag{5.3.29}
\end{aligned}
$$

Once $f_{R,m}$ is known we can compute $f_{L,m+1}$ by using (5.3.27), so we can focus on f_R and study the second order ODE (5.3.26). If we express this equation in terms of ψ and u we obtain the Schroedinger equation

$$\left(-\partial_u^2 + V\right)\psi = M_{F,m}^2 \psi, \tag{5.3.30}$$

where the "potential" V is given by

$$
\begin{aligned}
V(u) = & \ e\partial_u \mathcal{A}_\varphi e^{(A-B)/2} + \left(\frac{1}{2} + m + e\mathcal{A}_\varphi\right)\partial_u e^{(A-B)/2} \\
& + \left[\frac{1}{4} + m + e\mathcal{A}_\varphi + (m + e\mathcal{A}_\varphi)^2 \right] e^{A-B}. \tag{5.3.31}
\end{aligned}
$$

We observe that transformation (5.3.14) exactly removes the delta-functions which appear in (5.3.29) through Ω' (see Appendix C.2). However, just as for the gauge fields, a singular behaviour is observed in the potential, so that the asymptotic behaviour of the wave functions does not depend on their mass.

Our problem is now reduced to solving equation (5.3.30) with the conditions NC (5.3.13) and HC (5.3.15). By using (5.3.27) for $M_F \neq 0$ and definitions (4.4.40) and (5.3.14) we can rewrite (5.3.15) as follows

$$\left(\lim_{u\to\bar{u}} - \lim_{u\to 0}\right)\psi^* \left[-\partial_u + \left(m + \frac{1}{2} + e\mathcal{A}_\varphi\right)e^{(A-B)/2}\right]\psi = 0. \tag{5.3.32}$$

We can now proceed in exactly the same way as for the gauge field sector. For the conical-GGP solution (4.4.45) the explicit expression for V has the form (5.2.14), but now with

$$r_0^2 V_0 \equiv \left(\frac{1}{2} + m\right)\left[\bar{\varpi} - \omega + 2\omega\bar{\varpi}\left(\frac{1}{2} + m - N\right)\right] - \bar{\varpi}N, \quad (5.3.33)$$

$$r_0^2 v \equiv \left(\frac{1}{2} + m\right)\left[-\omega + \omega^2\left(\frac{1}{2} + m\right)\right], \quad (5.3.34)$$

$$r_0^2 \bar{v} \equiv \left(\frac{1}{2} + m\right)\left[\bar{\varpi} + \bar{\varpi}^2\left(\frac{1}{2} + m - 2N\right)\right]$$
$$+ \bar{\varpi}N\left(\bar{\varpi}N - 1\right). \quad (5.3.35)$$

Moreover, in this case the explicit expression for the HC is

$$\lim_{u \to \bar{u}} \psi^*\left(-\partial_u + \bar{\varpi}\frac{m + 1/2 - N}{\bar{u} - u}\right)\psi$$
$$- \lim_{u \to 0} \psi^*\left(-\partial_u + \omega\frac{m + 1/2}{u}\right)\psi = 0. \quad (5.3.36)$$

As in the gauge fields sector we introduce z and y in the following way

$$z = \cos^2\left(\frac{u}{r_0}\right), \qquad \psi = z^\gamma\left(1 - z\right)^\beta y(z), \quad (5.3.37)$$

so that equation (5.2.14) becomes a hypergeometric equation (5.2.18), with parameters:

$$\gamma \equiv \frac{1}{2}\left[1 + \bar{\varpi}\left(\frac{1}{2} + m - N\right)\right], \quad \beta \equiv \frac{1}{2}\left[1 - \omega\left(\frac{1}{2} + m\right)\right],$$

$$c \equiv \frac{3}{2} + \bar{\varpi}\left(\frac{1}{2} + m - N\right),$$

$$a \equiv 1 + \frac{\bar{\varpi}}{2}\left(\frac{1}{2} + m - N\right) - \frac{\omega}{2}\left(\frac{1}{2} + m\right) + \frac{1}{2}\sqrt{\Delta},$$

$$b \equiv 1 + \frac{\bar{\varpi}}{2}\left(\frac{1}{2} + m - N\right) - \frac{\omega}{2}\left(\frac{1}{2} + m\right) - \frac{1}{2}\sqrt{\Delta},$$

$$\Delta \equiv r_0^2 M_{F,m}^2 + (\bar{\varpi}N)^2 + \left(\frac{1}{2} + m\right)\left[\bar{\varpi}(\bar{\varpi} - 2\omega)\left(\frac{1}{2} + m - N\right)\right.$$
$$\left. - \bar{\varpi}^2 N + \omega^2\left(\frac{1}{2} + m\right)\right].$$

We can construct two independent solutions ψ_1 and ψ_2 of the Schroedinger equation (5.2.14) as in Section 5.2, and impose the NC (5.3.13) and

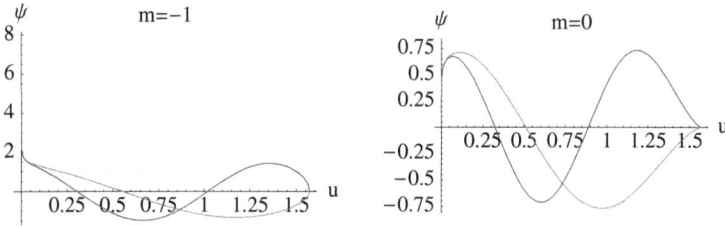

Figure 5.1: Fermion Wave Function Profiles: $n = 0, 1, 2$ modes plotted for angular momentum numbers $m = -1, 0$ (eqs (C.4.6) and (C.4.2) respectively). The parameters are chosen to be $(r_0, \omega, \overline{\omega}, e) = (1, 1/4, 1, 0)$, corresponding to a single negative tension brane. Also the normalisation constant is set to 1. The number of intersections with the u-axis equals n, according to quantum mechanics. Notice that the $(m, n) = (-1, 0)$ mode is massless, and that given a localized massless mode, there is also an infinite KK tower of localized massive modes.

the HC (5.3.36) to obtain the physical modes. The resulting wave functions are:

$$\psi = K_1\psi_1 + K_2\psi_2, \tag{5.3.38}$$

where the integration constants, $K_{1,2}$, are fixed in Appendix C.4. We plot a few of the wave function profiles in Figure 5.1.

The complete discrete mass spectrum is also given in Appendix C.4. There it can be seen that the same finiteness of the mass gap in the large α (hence large volume) limit, found in the gauge field spectrum, can be observed here. Moreover, for $\alpha \sim 1$, the mass gap between the zero modes and the massive states now goes as:

$$M_{GAP}^2 \sim \frac{1}{r_0^2} + \frac{1}{r_1^2}. \tag{5.3.39}$$

Therefore, for the fermions, a finite mass gap in the large volume limit can also be obtained by taking $r_0 \to \infty$ and turning on δ, thus allowing r_1 to remain finite. This contrasting behaviour to the standard KK picture is a consequence of the conical defects ($\overline{\omega} \neq 1$ and $\omega \neq 1$) in our internal manifold. Below we shall consider its implications for phenomenology.

5.3.3   4D Effective Fermion Charges

Let us first end this section on fermion fluctuations by obtaining their 4D effective gauge couplings. This can be calculated by going beyond their bilinear Lagrangian, and considering the interaction term:

$$\int d^6 X \sqrt{-G} \, \bar{\lambda} \Gamma^M \nabla_M \lambda = \cdots + \int d^6 X \sqrt{-G} \, \bar{\lambda} \Gamma^\mu (\partial_\mu + e V_\mu) \lambda + \dots \, .$$

$$(5.3.40)$$

Using the results for the KK decomposition found in the preceding sections:

$$\lambda(X) = \sum_{m,n} \lambda_{mn}(x) f_{mn}^{(\lambda)}(\rho) e^{im\varphi} = \sum_{m,n} \lambda_{mn}(x) \psi_{mn}^{(\lambda)}(u) e^{-A-B/4} e^{im\varphi}$$

$$V_\mu(X) = \sum_{m,n} V_{\mu \, mn}(x) f_{mn}^{(V)}(\rho) e^{im\varphi}$$

$$= \sum_{m,n} V_{\mu \, mn}(x) \psi_{mn}^{(V)}(u) e^{-(3A+B)/4} e^{im\varphi} \qquad (5.3.41)$$

and recalling that the gauge field zero mode is $f_{00}^{(V)} = 1$, a fermion mode λ_{mn} has the following coupling to the 4D effective gauge group :

$$
\begin{aligned}
e_{eff} &= \frac{e \int d\varphi du \, \overline{\psi_{mn}^{(\lambda)}} \, f_{00}^{(V)} \, \psi_{mn}^{(\lambda)}}{\int d\varphi du \, \overline{\psi_{mn}^{(\lambda)}} \, \psi_{mn}^{(\lambda)}} \\
&= e \qquad\qquad\qquad\qquad\qquad (5.3.42)
\end{aligned}
$$

Since the gauge field zero mode has a constant wave profile, the effective charges for the fermion modes are universal. This is a general result, and independent of any possible localization properties of the fermion modes: massless and massive fermion modes will always have the same coupling to the massless gauge fields. Again, we find that a large mass gap is required in the fermion spectrum in order to hide the KK tower. We now consider this issue in more detail.

5.4   Large Volume Compactifications with a Large Mass Gap

In the previous two sections, we have calculated the complete KK spectrum for the warped brane world compactification of 6D supergravity, for two interesting sectors of the gauge and matter fluctuations. We are now ready to consider the possible implications of our results.

6D brane world models have long been of interest in the context of

LED, since these may help with the gauge hierarchy problem. In the conventional ADD picture, Standard Model particles must be confined to a 4D brane world, in order to explain why the large extra dimensions have escaped detection. It would certainly be of interest to develop a dynamical description of this localization, within the context of low energy effective field theory.

This could be achieved, for instance, if the zero mode wave profiles were peaked near to a brane, and the heavy modes suppressed there [84]. However, we have found that zero mode fermions can be peaked near to negative tension branes, only at the price of localizing the whole KK tower (see Figure 5.1). Therefore, strong couplings are expected between light and heavy modes. If the zero mode bulk fermions are to be interpreted as matter in the SM, then apparently the only way to explain why we do not observe all the KK modes is by insisting that their mass gap is larger than the $100GeV$ scale probed to date.

Usually, this would bring us back to the classical KK scenario, with the extra dimensions required to be very small (at least $(100GeV)^{-1}$ scale), and the generation of the gauge hierarchy lost. However, in our framework we have seen that a large mass gap can occur even if the volume V_2, defined as the ratio κ^2/κ_4^2, becomes large. In the fermionic sector this set up can be achieved with the conical defect associated with the warping ($\omega \neq 1$ and $\overline{\omega} = 1$), by taking the parameter r_1 to be small. Another way is turning on the other defect ($\overline{\omega} \neq 1$) and then taking the large α limit, that is small $\overline{\omega}$ limit, corresponding to a negative tension brane. The volume V_2 in (5.1.5) becomes large but, for both the fermions and gauge fields, the mass gap does not reduce to zero. We observe that the latter mechanism works also for $r_0 = r_1$, that is $\omega = \overline{\omega}$, which corresponds to the unwarped "rugby ball" solution with branes of equal tension at each of the two poles [32, 34].

The general idea of relaxing the phenomenological constraints on the size of the extra dimensions by deforming the shape of the internal space was proposed in [87]. There, it was shown that the presence of shape moduli can imply that there is no experimental limit on the size of the largest extra dimension. However, requiring a large KK mass gap still constrained the overall volume of the extra dimensions. Here we give an explicit model which allows arbitrary large values for both V_2 and M_{GAP}^2, at least for the fermions and vectors. This could have an interesting application in the ADD scenario because we can have both $\kappa \sim TeV^{-2}$ and small effects from the massive modes by setting a large enough value of M_{GAP}.

In terms of hiding Large Extra Dimensions from our four dimensional universe, another possible approach is to interpret all the bulk fields that we have found (massless modes and massive ones) as a hidden sector, only gravitationally coupled to the SM. At this level, the SM must then be introduced by hand, confined on the delta-function brane. It seems that this is the approach to take if embedding the SLED scenario in our calculations, proposed in [55], to resolve the Cosmological Constant Problem. This proposal relates the hierarchy in the Electroweak scale with that of the Cosmological Constant. The Electroweak scale is set by the size of the extra dimensions, r, and the Cosmological Constant is given by the KK mass gap, here fixed by the same scale[12] $1/r$. Both may have their observed values when the 6D fundamental scale is TeV, and $r \sim 0.1mm$. For the mass gap to be this small, SM particles must be localized to the brane.

Let us end by considering the tunings involved, when constructing a model with large volume (say, $\sqrt{V_2} \sim 0.1mm$) and large mass gap (say, $M_{GAP} \sim TeV^{-1}$). Consider first a large mass gap for the fermions. If we set $\alpha \sim 1$ and $r_1 \ll r_0$, then the Dirac quantization (4.4.49) implies:

$$e\frac{r_1}{r_0}\frac{g}{g_1} \sim N \qquad (5.4.43)$$

If we then assume $e \sim 1$ (which is natural from group theory) and $N \sim 1$ (which is required for a small number of families), the large volume - large mass gap condition requires a large hierarchy in the bulk gauge couplings:

$$\frac{g}{g_1} \sim 10^{15}. \qquad (5.4.44)$$

Alternatively, we could set $r_1 \sim r_0$ and $\alpha \gg 1$, allowing a large mass gap for both fermions and gauge fields. In this case, requiring a large mass gap, $M_{GAP} \sim 1/r_0$, as well as large volume, $\sqrt{V_2} \sim \sqrt{\alpha}r_0$, requires $\alpha \sim 10^{30}$. Then, again, the Dirac quantization condition (4.4.49) reveals a large hierarchy in the bulk gauge couplings:

$$\frac{g}{g_1} \sim 10^{-30} \qquad (5.4.45)$$

In both scenarios we cannot embed the background monopole in $U(1)_R$.

---

[12]However, the breakdown of SUSY in the bulk, as in the solutions studied here, may lead to a larger prediction for the Cosmological Constant, see [55].

Gauge Fields	Fermions
$V \sim \quad (\mathbf{45}, \mathbf{1})_0$	$\lambda \sim \quad (\mathbf{45}, \mathbf{1})_1$
$+(\mathbf{16}, \mathbf{1})_0$	$+(\mathbf{16} + \overline{\mathbf{16}}, \mathbf{1})_1$
$+(\mathbf{1}, \mathbf{133})_0$	$+(\mathbf{1}, \mathbf{133})_1$
$+(\mathbf{1}, \mathbf{1})_0$	$\psi \sim (\mathbf{1}, \mathbf{912})_0$

Table 5.1: The gauge and fermion fields whose KK spectrum is given by our work, for the illustrative example of the anomaly free model $E_6 \times E_7 \times U(1)_R$, when the monopole is embedded in E_6. We give the quantum numbers under $\mathcal{H} = SO(10) \times E_7 \times U(1)_R$, which is the unbroken subgroup of the 6D gauge group.

There are of course other combinations, for example with both $\alpha \gg$ 1 and $r_1 \ll r_0$, in which these hierarchies may be relaxed. However, we should say that these tunings do not appear to be very natural or promising. For example, choosing the large dimensionless number $\alpha \gg 1$ corresponds to heavy negative tension branes, and deficit angles orders of magnitude less than zero. On the other hand, independently of trying to embed the Large Extra Dimension scenario into the present model, we have found an explicit example in which the KK mass gap does not go to zero as the volume goes to infinity, contrary to standard lore.

5.5 Conclusions and Outlook of Part II

In this part of the book we have analyzed an interesting subsector of gauge field and fermion fluctuations, in the warped brane world solutions of 6D minimal gauged supergravities. In particular, we have focused on bulk components which could give rise to SM or Grand Unified gauge and charged matter fields.

We performed a Fourier decomposition of 6D fields, and transformed the resulting field equations into a Schroedinger-like problem. We were then able to find the exact solutions for the KK modes, in terms of hypergeometric functions. We considered in detail the boundary conditions that the physical modes must satisfy. In addition to the normalizability constraint, consistency also required a hermiticity condition, which can be interpreted as demanding current conservation. We were able to implement this in its general, quadratic form. Together, these conditions selected the physical modes, and gave rise to a discrete mass spectrum, which we presented in full. The discreteness of the spectrum is of course to be expected given the compact topology of the internal manifold, whose Euler number is two.

Our study can be applied to several sectors of the 6D supergravities. In Table 5.1 we summarise the 6D fields that are covered by our analysis, for the illustrative example of the anomaly free model $E_6 \times E_7 \times U(1)$. Moreover, the corresponding spectra for the non-supersymmetric model of [88] (at least for the unwarped 4D Poincaré invariant case) and [32], generalized to Einstein-Yang-Mills with fermions, can be straightforwardly extracted from those given above simply by setting the warp factor to one, that is $r_1 = r_0$.

The exact results presented in this chapter enabled us to study the effects of the conical defects, sourced by codimension two branes, on the KK wave profiles and mass gaps. As usual, the gauge fields have a zero mode with constant wave profile. For the fermions, we found that some zero modes can be peaked on a negative tension brane, but in this case the whole KK tower is peaked there too. Therefore, in order to interpret the bulk zero modes as 4D effective fields of the SM, the mass gap must be large.

Intriguingly, this does not necessarily drive us to the conventional KK picture, with small compact dimensions. It does not, because the conical defects allow a novel behaviour in the mass gap, which can be decoupled from the volume of the compactification, defined by κ^2/κ_4^2. This continues to be observed in the unwarped limit, where the rugby ball model of [32, 34] is retrieved. A finite mass gap can be obtained, even if the volume goes to infinity.

For example, a large volume could be arranged in order to generate the Electroweak hierarchy, whilst maintaining a large mass gap between the zero modes and the KK tower. This picture does not seem to provide a realisation of the SLED scenario, where the volume and mass gap should be related. However, in this way, SM fields could arise from bulk fields, and along with gravity propagate through the large extra dimensions, perfectly consistently with observation. Moreover, for better or worse, this picture seems to render the LED scenario less falsifiable than previously thought, since we do not have to expect that the bulk KK modes are accessible at TeV scales. However, arranging for both a large volume and large mass gap seems to require a large degree of fine-tuning in bulk couplings. Furthermore, for a more complete idea, we would have to consider the KK spectrum for the remaining bulk sectors, and in particular the gravitational fluctuations to know the effect of LED on post-Newtonian tests[13].

Indeed, after the publication of the author's Ph.D. thesis, the analysis of the fluctuations presented in this part of the book was extended to other sectors. In Ref. [93] the perturbations coming from the internal components of the gauge fields orthogonal to the gauge field background were analysed, while Ref. [94, 95] extended the study to the full bosonic perturbations (at least in the simple unwarped case). A final objective would be to derive the full 4D effective field theory describing light fluctuations, and an understanding of when 6D physics comes into play. It should be possible to continue the project to other fields by extending our work, and find all KK modes and spectra including the fermions which mix with the graviton.

It is worth mentioning that the bosonic analysis of [94] was used in [96] to determine the gravitational contribution to modification to Newton's law due to the KK modes. This contribution, which was computed analytically, can be independent of the brane vacuum energy as a consequence of geometrical and topological properties of codimension two brane world studied in this part of the book. These results support the idea that in such models the gravitational interactions may be decoupled from the brane vacuum energy.

In another direction, much of our analysis was general, and could also be used to study other theories and other backgrounds with 4D Poincaré-2D axial symmetry.

Finally, it would be interesting to investigate whether there exist other mechanisms, which lead to the same decoupling between the mass gap and internal volume that we have found here. Indeed, in our set up we have been able to show that the decoupling arises due to the conical defects, but it may be possible to find other sources in different frameworks. In this way, our explicit example may be a realisation of a more general mechanism.

---

[13] Massless gravitational fluctuations in the non-supersymmetric rugby ball model were considered in [89] and, for thick branes, in [90]. Moreover, some results for the Kaluza-Klein spectra in non-supersymmetric warped brane models (with rugby ball limit) have been found in [91] and, for thick branes, in [92].

Further developments and Concluding Remarks

The 4D effective theory associated to a higher dimensional model must be derived carefully in order to obtain the correct physical predictions. In this book we have studied two important issues concerning the 4D interpretation of models with extra dimensions. The first one pertains to the role of heavy KK modes in the low energy dynamics. The second one is related to the dependence of the KK towers on possible conical defects of the internal manifold. In both cases we have analyzed examples which are interesting from the physical point of view. Indeed, in Chapter 3 we have studied a 6D non supersymmetric gauge and gravitational theory which leads to a 4D chiral effective theory similar to the electroweak part of the SM. Moreover, in Chapters 4 and 5 we have treated a 6D supergravity expanded around a non supersymmetric and singular solution, which could give rise to the SM or to a grand unified theory in the low energy limit. The latter calculation is also a first step towards the analysis of SLED as a scenario in which one can hope to solve the cosmological constant problem. Indeed, one can compute carefully all the contributions to the vacuum energy density only after constructing the complete 4D spectrum.

After the publication of the author's Ph.D. thesis, the Casimir energy due to bulk loops of matter fields in codimension-two brane worlds was studied in [97, 98, 99] and it was discussed how effective field theory methods allow us to use this result to renormalize the bulk and brane operators. In the calculation of [99] the sum over the Kaluza-Klein (KK) states was explicitly performed with a convenient method, which is based on a combined use of zeta function and dimensional regularization. Among the general class of models that were considered in [97, 98, 99] a supersymmetric example, 6D gauged chiral supergravity, was included. Although much of this discussion is more general, a class of compactifications was treated in some detail, where the extra dimensions parametrize a rugby ball shaped space with size stabilized by a bulk magnetic flux. The rugby ball geometry requires two branes, which can host the Standard Model fields and carry both tension and magnetic flux (of the bulk gauge field), the leading terms in a derivative expansion. The brane properties have an impact on the KK spectrum and therefore on the Casimir energy as well as on the renormalization of the brane operators. A very interesting feature is that when the two branes carry exactly the same amount of flux, one half of the bulk supersymmetries survives after the compactification, even if the brane tensions are large. It was also discussed the implications of these calculations for the natural value of the cosmological constant when the bulk has two large extra dimensions and the bulk supersymmetry is partially preserved (or completely broken).

Here we consider further possible outlooks which are shared by Parts I and II.

We observe that an interesting outlook could be the study of the scalar sector of 6D gauged supergravities expanded around an axisymmetric solution, for instance the conical-GGP solutions that we have presented in Subsection 4.4.2 and analyzed in Chapter 5. The contribution of the heavy KK modes to the scalar couplings may be non vanishing, as in the 6D Einstein-Maxwell-Scalar model, manifesting the underlying 6D physics. Moreover it would be interesting to know the form of the heavy mode contribution in the presence of warping and singularities in the internal manifold, which are both properties of the conical-GGP solutions.

Another possible outlook is the calculation of the contribution of the spin-0 fields to the modification of Newton's law.

Appendix A

Conventions and Notations

We choose the signature $-, +, +, +, \ldots$ for the metric G_{MN}. The Riemann tensor is defined as follows

$$R^R_{MNS} = \partial_M \Gamma^R_{NS} - \partial_N \Gamma^R_{MS} + \Gamma^R_{MP} \Gamma^P_{NS} - \Gamma^R_{NP} \Gamma^P_{MS}, \qquad \text{(A.0.1)}$$

where the $\Gamma's$ are the Levi-Civita connection. Whereas the Ricci tensor and the Ricci scalar

$$R_{MN} = R^P_{PMN}, \qquad R = G^{MN} R_{MN}. \qquad \text{(A.0.2)}$$

Here $M, N, \ldots$ run over all space-time dimensions.

Our choice for the 6D constant gamma matrices Γ^A, $A = 0, 1, 2, 3, 5, 6$, is

$$\Gamma^\mu = \begin{pmatrix} 0 & \gamma^\mu \\ \gamma^\mu & 0 \end{pmatrix}, \quad \Gamma^5 = \begin{pmatrix} 0 & \gamma^5 \\ \gamma^5 & 0 \end{pmatrix}, \quad \Gamma^6 = \begin{pmatrix} 0 & -i \\ i & 0 \end{pmatrix},$$
$$\text{(A.0.3)}$$

where the γ^μ are the 4D constant gamma matrices and γ^5 the 4D chirality matrix:

$$\gamma^5 = -i \gamma^0 \gamma^1 \gamma^2 \gamma^3. \qquad \text{(A.0.4)}$$

We define also the 6D chirality matrix Γ^7 by

$$\Gamma^7 = -\Gamma^0 \Gamma^1 \Gamma^2 \Gamma^3 \Gamma^5 \Gamma^6 = \begin{pmatrix} 1 & 0 \\ 0 & -1 \end{pmatrix}. \qquad \text{(A.0.5)}$$

Moreover the spin connection is

$$\omega_M^{[A,B]} = \eta^{BC}\omega_{M\ C}^{A} = \eta^{BC}\left(e_N^A \Gamma_{MR}^N e_B^R + e_N^A \partial_M e_B^N\right), \qquad (A.0.6)$$

where e_M^A is the vielbein. In the text e_M^A denotes the background spin connection and E_M^A the complete dynamical spin connection.

To study compactifications we split the D-dimensional space-time coordinates X^M, $M = 0, 1, ..., D - 1$ in two sets: 4D coordinates x^μ, $\mu = 0, 1, 2, 3$, and internal coordinates y^m, $m = 4, ..., D - 1$. A background metric with 4D Poincaré invariance reads

$$ds^2 = e^{A(y)}\eta_{\mu\nu}dx^\mu dx^\nu + g_{mn}(y)dy^m dy^n, \qquad (A.0.7)$$

where e^A is called warp factor and g_{mn} is the metric of the internal space.

We use the following further symbols.

$(Minkowski)_D$: D-dimensional Minkowski space-time.
d: number of extra dimensions ($d \equiv D - 4$).
K_d: internal d-dimensional space.
κ: D-dimensional Planck scale, the Einstein-Hilbert term in the lagrangian being R/κ.
κ_4: 4D Planck scale.
$\tilde{V}_d$: proper volume of K_d.
V_d: the ratio κ^2/κ_4^2; for unwarped geometries it equals $\tilde{V}_d$.
$r \equiv (V_d)^{1/d}$.
$\mathcal{M}^*$: complex conjugate of a matrix $\mathcal{M}$.
$\mathcal{O}^\dagger$: hermitian conjugate of an operator $\mathcal{O}$.
$< \mathcal{O} >$: vacuum expectation value of $\mathcal{O}$.
∇_M: covariant derivative, including gravitational and gauge connections.
a: radius of S^2.
$\delta^{(n)}(x)$: n-dimensional Dirac δ-function ($\int d^n x \delta^{(n)}(x) = 1$).

We also use the following abbreviations.

SM: Standard Model of Particle Physics.
GR: General Relativity.

KK: Kaluza-Klein.

VEV: Vacuum Expectation Value.

SSB: Spontaneous Symmetry Breaking.

EOM: Equation of Motion.

ADD: Arkani-Hamed, Dimopoulos and Dvali.

LED: Large Extra Dimensions.

RS: Randall-Sundrum

SLED: Supersymmetric Large Extra Dimensions.

Appendix B

Spectrum from 6D Einstein-Maxwell-Scalar Model

B.1 Spin-1 Mass Terms from S_F and S_R

B.1.1 S_F Contribution

In this subsection we write the contribution of

$$S_F \equiv -\frac{1}{4} \int d^6 X \sqrt{-G} F^2 \tag{B.1.1}$$

to the bilinear terms of V, U and W. By direct computation we get kinetic terms for V and U and some mass terms for U and W:

$$-\frac{1}{4} \int d^2 y \det\left(e_m^\alpha\right) F^2 = -\frac{1}{4} V_{\mu\nu} V^{\mu\nu} K - \frac{1}{6} U_{\mu\nu}^{\hat{\alpha}} U^{\mu\nu\hat{\beta}} K_{\hat{\alpha}\hat{\beta}}$$

$$-\frac{2}{3} U_\mu^{\hat{\alpha}} U^{\mu\hat{\beta}} M_{\hat{\alpha}\hat{\beta}}^{(1)} + \frac{4}{3} U_\mu^{\hat{\alpha}} W^{\mu\hat{\beta}} M_{\hat{\alpha}\hat{\beta}}^{(2)} - \frac{2}{3} W_\mu^{\hat{\alpha}} W^{\mu\hat{\beta}} M_{\hat{\alpha}\hat{\beta}}^{(3)} + ..., \tag{B.1.2}$$

where the 4D curved indices μ and ν are contracted with the 4D metric $g_{\mu\nu}$, the dots are constant terms and interaction terms, moreover

$$V_{\mu\nu} = \partial_\mu V_\nu - \partial_\nu V_\mu, \quad U_{\mu\nu}^{\hat{\alpha}} = \partial_\mu U_\nu^{\hat{\alpha}} - \partial_\nu U_\mu^{\hat{\alpha}} \tag{B.1.3}$$

and

$$K = \frac{1}{4\pi a^2} \int d^2y \det\left(e_m^\alpha\right),$$

$$K_{\hat\alpha\hat\beta} = \frac{3}{4\pi} \left(\frac{\kappa}{\sqrt{2}ea^2}\right)^2 \int d^2y \det\left(e_m^\alpha\right) \mathcal{D}_{\hat\alpha}^3 \mathcal{D}_{\hat\beta}^3,$$

$$M_{\hat\alpha\hat\beta}^{(1)} = \frac{3}{8\pi} \left(\frac{\kappa}{\sqrt{2}ea^2}\right)^2 \int d^2y \det\left(e_m^\alpha\right) g^{mn} \partial_m \mathcal{D}_{\hat\alpha}^3 \partial_n \mathcal{D}_{\hat\beta}^3$$

$$M_{\hat\alpha\hat\beta}^{(2)} = -\frac{3\kappa^2}{16\pi ea^3} \int d^2y \det\left(e_m^\alpha\right) \partial_m \mathcal{D}_{\hat\alpha}^3 \mathcal{D}_{\hat\beta}^3 e_\alpha^n g^{mq} F_{nq}$$

$$M_{\hat\alpha\hat\beta}^{(3)} = \frac{3\kappa^2}{16\pi a^2} \int d^2y \det\left(e_m^\alpha\right) \mathcal{D}_{\hat\alpha}^\alpha e_\alpha^m \mathcal{D}_{\hat\beta}^\beta e_\beta^p F_p{}^n F_{mn}. \tag{B.1.4}$$

The results (B.1.4) are valid for all background e^α and e^3. We use the $SU(2) \times U(1)$ background in the Subsection B.1.3, the $U(1)_3$ background in the subsection B.1.4.

B.1.2 S_R Contribution

In this subsection we write the contribution of

$$S_R \equiv \int d^6X \sqrt{-G} \frac{1}{\kappa^2} R \tag{B.1.5}$$

to the bilinear terms of W. The complete contribution of S_R to the 4D action is given in [5] in the case of non deformed background solutions. Here we need explicit expressions, at least for the bilinears, which are also valid for deformed solutions. We get a kinetic term and a mass term of W: up to a total derivative we have

$$\int d^2y \frac{1}{\kappa^2} \det\left(e_m^\alpha\right) R = -\frac{1}{6} W_{\mu\nu}^{\hat\alpha} W^{\mu\nu\hat\beta} K'_{\hat\alpha\hat\beta}$$
$$+ W_\mu^{\hat\alpha} W^{\mu\hat\beta} M_{\hat\alpha\hat\beta}^{(4)} + ..., \tag{B.1.6}$$

where the dots include constant and interaction terms; moreover

$$W_{\mu\nu}^{\hat\alpha} = \partial_\mu W_\nu^{\hat\alpha} - \partial_\nu W_\mu^{\hat\alpha}, \tag{B.1.7}$$

and

$$K'_{\hat{\alpha}\hat{\beta}} = \frac{3}{8\pi a^2} \int d^2y \det(e^{\alpha}_m) \, \mathcal{D}^{\alpha}_{\hat{\alpha}}\mathcal{D}^{\beta}_{\hat{\beta}} g_{\alpha\beta},$$

$$\begin{aligned}
M^{(4)}_{\hat{\alpha}\hat{\beta}} = \frac{1}{4\pi a^2} \int d^2y \det(e^{\alpha}_m) \Bigg[& \partial_n \mathcal{D}^{\alpha}_{\hat{\alpha}}\mathcal{D}^{\beta}_{\hat{\beta}} \big(-e^m_{\alpha} \omega_m{}^{\gamma}{}_{\beta} e^n_{\gamma} \\
& - g_{\alpha\delta} g^{nm} \omega_m{}^{\delta}{}_{\beta} + 2e^n_{\alpha}e^m_{\gamma} \omega_m{}^{\gamma}{}_{\beta} \big) \\
& + \mathcal{D}^{\alpha}_{\hat{\alpha}}\mathcal{D}^{\beta}_{\hat{\beta}} \left(-\frac{1}{2}\omega_n{}^{\gamma}{}_{\alpha} e^m_{\gamma} \omega_m{}^{\delta}{}_{\beta} e^n_{\delta} - \frac{1}{2}\omega_n{}^{\delta}{}_{\alpha} g^{nm} \omega_{m\delta\beta} + \omega_n{}^{\delta}{}_{\alpha} e^n_{\delta} \omega_m{}^{\gamma}{}_{\beta} e^m_{\gamma} \right) \\
& + \partial_n \mathcal{D}^{\alpha}_{\hat{\alpha}} \partial_m \mathcal{D}^{\beta}_{\hat{\beta}} \left(-\frac{1}{2}e^m_{\alpha}e^n_{\beta} + e^n_{\alpha}e^m_{\beta} - \frac{1}{2}g_{\alpha\beta}g^{nm} \right) \Bigg],
\end{aligned} \tag{B.1.8}$$

where $\omega_n{}^{\alpha}{}_{\beta}$ is the 2-dimensional spin connection for e^{α}_n. The results (B.1.8) are also valid for every background e^{α} and e^3. We use the $SU(2) \times U(1)$ background in the Subsection B.1.3, the $U(1)_3$ background in the Subsection B.1.4.

B.1.3   The Case of $SU(2) \times U(1)$ background

We use now the $SU(2) \times U(1)$ background, that is $\eta = 0$. This computation is performed in [7]. We have the following bilinear terms for V, U and W:

$$-\frac{1}{4}V_{\mu\nu}V^{\mu\nu} - \frac{1}{6}U^{\hat{\alpha}}_{\mu\nu}U^{\mu\nu}_{\hat{\alpha}} - \frac{1}{6}W^{\hat{\alpha}}_{\mu\nu}W^{\mu\nu}_{\hat{\alpha}}$$
$$-\frac{2}{3a^2}(U_{\mu\hat{\alpha}} - W_{\mu\hat{\alpha}})\left(U^{\mu\hat{\alpha}} - W^{\mu\hat{\alpha}}\right). \tag{B.1.9}$$

If we define

$$\begin{aligned}
\mathcal{A} &= \sqrt{\frac{1}{3}}(W + U), \\
X &= \sqrt{\frac{1}{3}}(W - U),
\end{aligned} \tag{B.1.10}$$

we can write (B.1.9) as follows

$$-\frac{1}{4}V_{\mu\nu}V^{\mu\nu} - \frac{1}{4}\mathcal{A}^{\hat{\alpha}}_{\mu\nu}\mathcal{A}^{\mu\nu}_{\hat{\alpha}}$$
$$-\frac{1}{4}X^{\hat{\alpha}}_{\mu\nu}X^{\mu\nu}_{\hat{\alpha}} - \frac{2}{a^2}X_{\mu\hat{\alpha}}X^{\mu\hat{\alpha}}, \tag{B.1.11}$$

So $\mathcal{A}$ is a massless field, in fact it's the $SU(2)$ Yang-Mills field [7], while X is a massive field which can be neglected in the low energy limit.

B.1.4   The Case of $U(1)_3$ Background

Let us consider now the solution (3.3.12). First we note that S_R and S_F do not give mass terms for V; so the only source for the mass of V is S_ϕ.

We want to prove now that also the $SU(2)$ Yang-Mills fields masses do not receive contributions from S_R and S_F. First we give the bilinears for U and W, which come from S_R and S_F:

$$-\frac{1}{6}U^{\hat\alpha}_{\mu\nu}U^{\mu\nu\hat\beta}g_{\hat\alpha\hat\beta}\left(1+|\eta|\beta k_{\hat\alpha}\right)-\frac{1}{6}W^{\hat\alpha}_{\mu\nu}W^{\mu\nu\hat\beta}g_{\hat\alpha\hat\beta}\left(1+|\eta|\beta k'_{\hat\alpha}\right)$$

$$-\frac{2}{3}U^{\hat\alpha}_{\mu}U^{\mu\hat\beta}g_{\hat\alpha\hat\beta}\left(1+|\eta|\beta m^{(1)}_{\hat\alpha}\right)+\frac{4}{3}U^{\hat\alpha}_{\mu}W^{\mu\hat\beta}g_{\hat\alpha\hat\beta}\left(1+|\eta|\beta m^{(2)}_{\hat\alpha}\right)$$

$$-\frac{2}{3}W^{\hat\alpha}_{\mu}W^{\mu\hat\beta}g_{\hat\alpha\hat\beta}\left(1+|\eta|\beta m^{(3)}_{\hat\alpha}\right),$$

$$(B.1.12)$$

where

$$k_+ = k_- = \frac{2}{5},\ \ k_3 = \frac{1}{5},\ \ k'_+ = k'_- = \frac{3}{10},\ \ k'_3 = \frac{2}{5},$$

$$m^{(1)}_+ = m^{(1)}_- = \frac{1}{5},\ \ m^{(1)}_3 = -\frac{2}{5},$$

$$m^{(2)}_+ = m^{(2)}_- = -\frac{1}{20},\ \ m^{(2)}_3 = -\frac{2}{5},\ \ m^{(3)}_+ = m^{(3)}_- = -\frac{3}{10},\ \ m^{(3)}_3 = -\frac{2}{5}.$$

In order to prove (B.1.12) it's useful to use the following formula for the background spin connection:

$$\omega^{+}_{\varphi\ +} = -\omega^{-}_{\varphi\ -} = \frac{i}{a}\left(\cos\theta-1-\frac{1}{2}|\eta|\beta\cos\theta\sin^2\theta\right)$$

$$(B.1.13)$$

and $\omega^{-}_{\varphi\ +} = \omega^{+}_{\varphi\ -} = 0$.

Now we define X and A as follows

$$\left(1+\frac{|\eta|\beta}{2}k'_{\hat\alpha}\right)W^{\hat\alpha} = \sqrt{\frac{3}{2}}\left(\cos\theta^{\hat\alpha}_\eta X^{\hat\alpha}+\sin\theta^{\hat\alpha}_\eta \mathcal{A}^{\hat\alpha}\right),$$

$$\left(1+\frac{|\eta|\beta}{2}k_{\hat\alpha}\right)U^{\hat\alpha} = \sqrt{\frac{3}{2}}\left(-\sin\theta^{\hat\alpha}_\eta X^{\hat\alpha}+\cos\theta^{\hat\alpha}_\eta \mathcal{A}^{\hat\alpha}\right),$$

$$(B.1.14)$$

where the angle $\theta^{\hat\alpha}_\eta$ is defined by

$$\cos\theta^{\hat\alpha}_\eta = \frac{1+|\eta|\beta\delta^{\hat\alpha}}{\sqrt{2}},\ \ \sin\theta^{\hat\alpha}_\eta = \frac{1-|\eta|\beta\delta^{\hat\alpha}}{\sqrt{2}},$$

$$(B.1.15)$$

and the quantities $\delta^{\hat\alpha}$ are not still fixed. It's simple to check that the kinetic terms for X and $\mathcal{A}$ are in the standard form for every $\delta^{\hat\alpha}$ up to $O(\eta^{3/2})$. The definition (B.1.14) reduce to (B.1.10) for $\eta = 0$.

If we choose

$$\delta^{\hat\alpha} = \frac{1}{8}\left(m^{(3)}_{\hat\alpha}-k'_{\hat\alpha}-m^{(1)}_{\hat\alpha}+k_{\hat\alpha}\right)$$

$$(B.1.16)$$

we have no mass terms for $\mathcal{A}$ coming from $S_R + S_F$.

So the only source for the spin-1 low energy spectrum is S_ϕ and the result is given in equations (3.3.21) and (3.3.22).

B.2 Explicit Calculation of Spin-0 Spectrum

As we pointed out in the text, in order to find the spin-0 spectrum the expression of the $|i>$, $i = 1, ..., 6$, vectors is needed; these are defined by $\mathcal{O}_0|i> = 0$, which is equivalent to $\nabla^2\phi + \phi/a^2 = 0$, where $\nabla^2\phi$ is the Laplacian over the charged scalar ϕ, calculated with the round S^2 metric. Our choice for the orthonormal vectors[1] $|i>$ is

$$
|1> = \frac{1}{\sqrt{2}}\begin{pmatrix} \sqrt{\frac{3}{4\pi}}\mathcal{D}^{(1)}_{-1,1} \\ \sqrt{\frac{3}{4\pi}}\left(\mathcal{D}^{(1)}_{-1,1}\right)^* \\ 0 \\ \cdot \\ \cdot \\ \cdot \\ 0 \end{pmatrix}, \quad |2> = \frac{1}{\sqrt{2}}\begin{pmatrix} i\sqrt{\frac{3}{4\pi}}\mathcal{D}^{(1)}_{-1,1} \\ -i\sqrt{\frac{3}{4\pi}}\left(\mathcal{D}^{(1)}_{-1,1}\right)^* \\ 0 \\ \cdot \\ \cdot \\ \cdot \\ 0 \end{pmatrix},
$$

$$
|3> = \frac{1}{\sqrt{2}}\begin{pmatrix} \sqrt{\frac{3}{4\pi}}\mathcal{D}^{(1)}_{-1,0} \\ \sqrt{\frac{3}{4\pi}}\left(\mathcal{D}^{(1)}_{-1,0}\right)^* \\ 0 \\ \cdot \\ \cdot \\ \cdot \\ 0 \end{pmatrix}, \quad |4> = \frac{1}{\sqrt{2}}\begin{pmatrix} i\sqrt{\frac{3}{4\pi}}\mathcal{D}^{(1)}_{-1,0} \\ -i\sqrt{\frac{3}{4\pi}}\left(\mathcal{D}^{(1)}_{-1,0}\right)^* \\ 0 \\ \cdot \\ \cdot \\ \cdot \\ 0 \end{pmatrix},
$$

$$
|5> = \frac{1}{\sqrt{2}}\begin{pmatrix} \sqrt{\frac{3}{4\pi}}\mathcal{D}^{(1)}_{-1,-1} \\ \sqrt{\frac{3}{4\pi}}\left(\mathcal{D}^{(1)}_{-1,-1}\right)^* \\ 0 \\ \cdot \\ \cdot \\ \cdot \\ 0 \end{pmatrix}, \quad |6> = \frac{1}{\sqrt{2}}\begin{pmatrix} i\sqrt{\frac{3}{4\pi}}\mathcal{D}^{(1)}_{-1,-1} \\ -i\sqrt{\frac{3}{4\pi}}\left(\mathcal{D}^{(1)}_{-1,-1}\right)^* \\ 0 \\ \cdot \\ \cdot \\ \cdot \\ 0 \end{pmatrix}.
$$

Another ingredient for the calculation of the spin-0 spectrum is an explicit expression of the vectors $|\tilde{i}>$ and of the eigenvalues $M^2_{\tilde{i}}$; the latter are

---

[1] We express a generic vector as in (3.3.36).

given in [100, 101], however, we need here also a correspondence between eigenvalues and eigenvectors. As we explained in the text, only the $|\tilde{i}>$ like (3.3.41) and made of $l = 0, 1, 2$ harmonics are needed. The $|\tilde{i}>$ vectors must satisfy the following eigenvalue equations[2]:

$$-\nabla^2 h_{++} + 2R_{+-+-}h_{++} - 2\kappa^2 F_{+-}^2 h_{++} - \sqrt{2}\kappa\nabla_+\mathcal{V}_+F_{-+} = M^2 h_{++},$$

$$-\nabla^2 h_{--} + 2R_{+-+-}h_{--} - 2\kappa^2 F_{+-}^2 h_{--} + \sqrt{2}\kappa\nabla_-\mathcal{V}_-F_{-+} = M^2 h_{--},$$

$$-\nabla^2 h_{+-} - R_{+-+-}h_{+-} - \frac{\kappa}{\sqrt{2}}\nabla_+\mathcal{V}_-F_{-+} + \frac{\kappa}{\sqrt{2}}\nabla_-\mathcal{V}_+F_{-+} = M^2 h_{+-},$$

$$-\nabla^2 \mathcal{V}_+ + R_{+-}\mathcal{V}_+ - \kappa^2\mathcal{V}_+F_{+-}^2 + \frac{\kappa}{\sqrt{2}}\nabla_+h_{+-}F_{-+} - \sqrt{2}\kappa\nabla_-h_{++}F_{-+}$$

$$= M^2\mathcal{V}_+,$$

$$-\nabla^2\mathcal{V}_- + R_{+-}\mathcal{V}_- - \kappa^2\mathcal{V}_-F_{+-}^2 - \frac{\kappa}{\sqrt{2}}\nabla_-h_{+-}F_{-+} + \sqrt{2}\kappa\nabla_+h_{--}F_{-+}$$

$$= M^2\mathcal{V}_-, \tag{B.2.1}$$

where the background objects $(\nabla^2, R_{+-+-}, ...)$ correspond to the background (3.2.1), (3.2.2) and (3.2.3). We can transform the differential problem (B.2.1) into an algebraic one by using the expansion (3.2.11). We get an eigenvalue problem for every value of l and we give now an explicit expression for the $|\tilde{i}>$ vectors for the relevant value of l, namely $l = 0, 1, 2$. For $l = 0$ we get just one eigenvector $|\tilde{1}>$ with $M^2 = 1/a^2$:

$$|\tilde{1}> = \begin{pmatrix} 0 \\ 0 \\ 0 \\ 0 \\ 1/\sqrt{4\pi} \\ 0 \\ 0 \end{pmatrix}. \tag{B.2.2}$$

For $l = 1$ we get three different eigenvalues: $M^2 = 2/a^2, 4/a^2, 5/a^2$. The eigenvectors which correspond to $M^2 = 2/a^2$ are

$$|\tilde{2}_0 >, \quad \frac{1}{\sqrt{2}}\left(|\tilde{2}_1 > +|\tilde{2}_{-1} >\right), \quad \frac{1}{\sqrt{2i}}\left(|\tilde{2}_1 > -|\tilde{2}_{-1} >\right), \tag{B.2.3}$$

---

[2] We derive (B.2.1) evaluating (3.3.30), (3.3.31) and (3.3.34) in the basis (3.2.6) and performing the redefinition $h_{\pm\pm} \rightarrow \sqrt{2}\kappa h_{\pm\pm}$ and $h_{+-} \rightarrow h_{+-}\kappa/\sqrt{2}$, which normalizes the kinetic terms in the standard way.

where

$$|\tilde{2}_m> \equiv \frac{1}{\sqrt{6}} \begin{pmatrix} 0 \\ 0 \\ 0 \\ 0 \\ 2\sqrt{\frac{3}{4\pi}}\mathcal{D}_{0,m}^{(1)} \\ -\sqrt{\frac{3}{4\pi}}\mathcal{D}_{1,m}^{(1)} \\ -\sqrt{\frac{3}{4\pi}}\mathcal{D}_{-1,m}^{(1)} \end{pmatrix} . \tag{B.2.4}$$

Instead the eigenvectors which correspond to $M^2 = 4/a^2$ are

$$i|\tilde{3}_0>, \quad \frac{1}{\sqrt{2i}}\left(|\tilde{3}_1>+|\tilde{3}_{-1}>\right), \quad \frac{1}{\sqrt{2}}\left(|\tilde{3}_1>-|\tilde{3}_{-1}>\right), \tag{B.2.5}$$

where

$$|\tilde{3}_m> \equiv \frac{1}{\sqrt{2}} \begin{pmatrix} 0 \\ 0 \\ 0 \\ 0 \\ 0 \\ -\sqrt{\frac{3}{4\pi}}\mathcal{D}_{1,m}^{(1)} \\ \sqrt{\frac{3}{4\pi}}\mathcal{D}_{-1,m}^{(1)} \end{pmatrix} . \tag{B.2.6}$$

Moreover the eigenvectors which correspond to $M^2 = 5/a^2$ are

$$|\tilde{4}_0>, \quad \frac{1}{\sqrt{2}}\left(|\tilde{4}_1>+|\tilde{4}_{-1}>\right), \quad \frac{1}{\sqrt{2i}}\left(|\tilde{4}_1>-|\tilde{4}_{-1}>\right), \tag{B.2.7}$$

where

$$|\tilde{4}_m> \equiv \frac{1}{\sqrt{3}} \begin{pmatrix} 0 \\ 0 \\ 0 \\ 0 \\ \sqrt{\frac{3}{4\pi}}\mathcal{D}_{0,m}^{(1)} \\ \sqrt{\frac{3}{4\pi}}\mathcal{D}_{1,m}^{(1)} \\ \sqrt{\frac{3}{4\pi}}\mathcal{D}_{-1,m}^{(1)} \end{pmatrix} . \tag{B.2.8}$$

Finally, for $l = 2$ the values of M^2 are given by

$$a^2 M^2 = 6, \ 2(3-\sqrt{3}), \ 2(3+\sqrt{3}), \ \frac{1}{2}(13-\sqrt{73}), \ \frac{1}{2}(13+\sqrt{73}). \tag{B.2.9}$$

The eigenvectors with $a^2 M^2 = 6$ are

$$|\tilde{5}_0>, \quad \frac{1}{\sqrt{2}}\left(|\tilde{5}_1>-|\tilde{5}_{-1}>\right), \quad \frac{1}{\sqrt{2i}}\left(|\tilde{5}_1>+|\tilde{5}_{-1}>\right),$$

$$\frac{1}{\sqrt{2}}\left(|\tilde{5}_2>+|\tilde{5}_{-2}>\right), \quad \frac{1}{\sqrt{2i}}\left(|\tilde{5}_2>-|\tilde{5}_{-2}>\right), \tag{B.2.10}$$

where

$$|\tilde{5}_m> \equiv \frac{1}{3\sqrt{2}}\begin{pmatrix} 0 \\ 0 \\ -\sqrt{2}\sqrt{\frac{5}{4\pi}}\mathcal{D}^{(2)}_{2,m} \\ -\sqrt{2}\sqrt{\frac{5}{4\pi}}\mathcal{D}^{(2)}_{-2,m} \\ -2\sqrt{3}\sqrt{\frac{5}{4\pi}}\mathcal{D}^{(2)}_{0,m} \\ -\sqrt{\frac{5}{4\pi}}\mathcal{D}^{(2)}_{1,m} \\ \sqrt{\frac{5}{4\pi}}\mathcal{D}^{(2)}_{-1,m} \end{pmatrix}. \tag{B.2.11}$$

For $a^2 M^2 = 2(3-\sqrt{3})$ we have the eigenvectors

$$i|\tilde{6}_0>, \quad \frac{1}{\sqrt{2}}\left(|\tilde{6}_1>+|\tilde{6}_{-1}>\right), \quad \frac{1}{\sqrt{2i}}\left(|\tilde{6}_1>-|\tilde{6}_{-1}>\right),$$

$$\frac{1}{\sqrt{2}}\left(|\tilde{6}_2>-|\tilde{6}_{-2}>\right), \quad \frac{1}{\sqrt{2i}}\left(|\tilde{6}_2>+|\tilde{6}_{-2}>\right), \tag{B.2.12}$$

where

$$|\tilde{6}_m> \equiv \frac{1}{\sqrt{2(3+\sqrt{3})}}\begin{pmatrix} 0 \\ 0 \\ -\frac{1+\sqrt{3}}{\sqrt{2}}\sqrt{\frac{5}{4\pi}}\mathcal{D}^{(2)}_{2,m} \\ \frac{1+\sqrt{3}}{\sqrt{2}}\sqrt{\frac{5}{4\pi}}\mathcal{D}^{(2)}_{-2,m} \\ 0 \\ \sqrt{\frac{5}{4\pi}}\mathcal{D}^{(2)}_{1,m} \\ \sqrt{\frac{5}{4\pi}}\mathcal{D}^{(2)}_{-1,m} \end{pmatrix}. \tag{B.2.13}$$

For $a^2 M^2 = 2(3+\sqrt{3})$ we have the eigenvectors

$$i|\tilde{7}_0>, \quad \frac{1}{\sqrt{2}}\left(|\tilde{7}_1>+|\tilde{7}_{-1}>\right), \quad \frac{1}{\sqrt{2i}}\left(|\tilde{7}_1>-|\tilde{7}_{-1}>\right),$$

$$\frac{1}{\sqrt{2}}\left(|\tilde{7}_2>-|\tilde{7}_{-2}>\right), \quad \frac{1}{\sqrt{2i}}\left(|\tilde{7}_2>+|\tilde{7}_{-2}>\right), \tag{B.2.14}$$

where

$$|\tilde{7}_m> \equiv \frac{1}{\sqrt{2(3-\sqrt{3})}} \begin{pmatrix} 0 \\ 0 \\ -\frac{1-\sqrt{3}}{\sqrt{2}}\sqrt{\frac{5}{4\pi}}\mathcal{D}^{(2)}_{2,m} \\ \frac{1-\sqrt{3}}{\sqrt{2}}\sqrt{\frac{5}{4\pi}}\mathcal{D}^{(2)}_{-2,m} \\ 0 \\ \sqrt{\frac{5}{4\pi}}\mathcal{D}^{(2)}_{1,m} \\ \sqrt{\frac{5}{4\pi}}\mathcal{D}^{(2)}_{-1,m} \end{pmatrix}. \qquad (B.2.15)$$

Then for $a^2 M^2 = (13 - \sqrt{73})/2$:

$$|\tilde{8}_0>, \quad \frac{1}{\sqrt{2}}\left(|\tilde{8}_1>-|\tilde{8}_{-1}>\right), \quad \frac{1}{\sqrt{2i}}\left(|\tilde{8}_1>+|\tilde{8}_{-1}>\right),$$

$$\frac{1}{\sqrt{2}}\left(|\tilde{8}_2>+|\tilde{8}_{-2}>\right), \quad \frac{1}{\sqrt{2i}}\left(|\tilde{8}_2>-|\tilde{8}_{-2}>\right), \qquad (B.2.16)$$

where

$$|\tilde{8}_m> \equiv \frac{1+\sqrt{73}}{\sqrt{438+30\sqrt{73}}} \begin{pmatrix} 0 \\ 0 \\ \frac{13\sqrt{2}+\sqrt{146}}{2(1+\sqrt{73})}\sqrt{\frac{5}{4\pi}}\mathcal{D}^{(2)}_{2,m} \\ \frac{13\sqrt{2}+\sqrt{146}}{2(1+\sqrt{73})}\sqrt{\frac{5}{4\pi}}\mathcal{D}^{(2)}_{-2,m} \\ -\frac{4\sqrt{3}}{1+\sqrt{73}}\sqrt{\frac{5}{4\pi}}\mathcal{D}^{(2)}_{0,m} \\ -\sqrt{\frac{5}{4\pi}}\mathcal{D}^{(2)}_{1,m} \\ \sqrt{\frac{5}{4\pi}}\mathcal{D}^{(2)}_{-1,m} \end{pmatrix}. \qquad (B.2.17)$$

Finally for $a^2 M^2 = (13 + \sqrt{73})/2$:

$$|\tilde{9}_0>, \quad \frac{1}{\sqrt{2}}\left(|\tilde{9}_1>-|\tilde{9}_{-1}>\right), \quad \frac{1}{\sqrt{2i}}\left(|\tilde{9}_1>+|\tilde{9}_{-1}>\right),$$

$$\frac{1}{\sqrt{2}}\left(|\tilde{9}_2>+|\tilde{9}_{-2}>\right), \quad \frac{1}{\sqrt{2i}}\left(|\tilde{9}_2>-|\tilde{9}_{-2}>\right), \qquad (B.2.18)$$

where

$$|\tilde{9}_m> \equiv \frac{1-\sqrt{73}}{\sqrt{438-30\sqrt{73}}} \begin{pmatrix} 0 \\ 0 \\ \frac{13\sqrt{2}-\sqrt{146}}{2(1-\sqrt{73})}\sqrt{\frac{5}{4\pi}}\mathcal{D}^{(2)}_{2,m} \\ \frac{13\sqrt{2}-\sqrt{146}}{2(1-\sqrt{73})}\sqrt{\frac{5}{4\pi}}\mathcal{D}^{(2)}_{-2,m} \\ -\frac{4\sqrt{3}}{1-\sqrt{73}}\sqrt{\frac{5}{4\pi}}\mathcal{D}^{(2)}_{0,m} \\ -\sqrt{\frac{5}{4\pi}}\mathcal{D}^{(2)}_{1,m} \\ \sqrt{\frac{5}{4\pi}}\mathcal{D}^{(2)}_{-1,m} \end{pmatrix}. \tag{B.2.19}$$

We can now calculate the 6×6 matrix M^2_{ij} given in (3.3.39). In order to do that we need just the matrix elements $<i|\mathcal{O}_1|\tilde{i}>$ and $<i|\mathcal{O}_2|j>$, which can be computed by evaluating[3] $\mathcal{L}_0(\phi,h)$ and $\mathcal{L}_0(\phi,\phi)$, which appears in (3.3.29) and (3.3.32), in the $\pm$ basis given in (3.2.6). After the redefinitions $h_{\pm\pm}\to\sqrt{2}kh_{\pm\pm}$ and $h_{+-}\to h_{+-}k/\sqrt{2}$, which normalize the kinetic terms in the standard way, we get (for $n=2$)

$$\begin{aligned} \mathcal{L}_0(\phi,h) &= \sqrt{2}\kappa\nabla_+\Phi\nabla_+h_{--}\phi^* + \frac{\kappa}{\sqrt{2}}\nabla_+\Phi\nabla_-h_{+-}\phi^* \\ &\quad + \sqrt{2}\kappa\nabla_+\Phi h_{--}\left(\nabla_-\phi\right)^* + \frac{\kappa}{\sqrt{2}}\nabla_+\Phi h_{+-}\left(\nabla_+\phi\right)^* + c.c.\,, \\ \mathcal{L}_0(\phi,\phi) &= \phi^*\partial^2\phi - \phi^*\left[-\nabla^2 + m^2 + (e^2+4\xi)\Phi^*\Phi + \kappa^2\nabla_+\Phi\left(\nabla_+\Phi\right)^*\right]\phi \\ &\quad - \frac{1}{2}\left[\phi(2\xi - e^2)\left(\Phi^*\right)^2\phi + c.c.\right]. \end{aligned} \tag{B.2.20}$$

By using these expressions and the values of $|i>$ and $|\tilde{i}>$ given before, we find the following expression for M^2_{ij}:

$$\{M^2_{ij}\} = \begin{pmatrix} a_1 & 0 & 0 & 0 & a_4 & 0 \\ 0 & a_1 & 0 & 0 & 0 & -a_4 \\ 0 & 0 & a_2 & 0 & 0 & 0 \\ 0 & 0 & 0 & a_3 & 0 & 0 \\ a_4 & 0 & 0 & 0 & a_1 & 0 \\ 0 & -a_4 & 0 & 0 & 0 & a_1 \end{pmatrix}, \tag{B.2.21}$$

---

[3]For the background solution (3.2.1), (3.2.2) and (3.2.3) we have $\mathcal{L}_0(\phi,\mathcal{V})=0$.

where

$$a_1 = \frac{|\eta|}{a^2}\left(-sign(\eta) + \frac{3}{10}\beta + \frac{12}{5}\frac{\beta\xi a^2}{\kappa^2}\right),$$

$$a_2 = \frac{|\eta|}{a^2}\left(-sign(\eta) - \frac{6}{5}\beta + \frac{24}{5}\frac{\beta\xi a^2}{\kappa^2}\right),$$

$$a_3 = \frac{|\eta|}{a^2}\left(-sign(\eta) + \frac{4}{15}\beta + \frac{8}{5}\frac{\beta\xi a^2}{\kappa^2}\right),$$

$$a_4 = \frac{|\eta|}{a^2}\beta\left(\frac{3}{10} - \frac{4}{5}\frac{\xi a^2}{\kappa^2}\right). \tag{B.2.22}$$

By diagonalizing M_{ij}^2, we found exactly the spectrum that we discussed in the Subsection 3.3.2: the squared masses of the vector particles are reproduced[4], as required by the light cone gauge; moreover we get the two masses squared given in (3.3.42).

B.3 Explicit Calculation of Spin-1/2 Spectrum

Here we concentrate on the right-handed sector, which is the non trivial one because it presents $\eta^{1/2}$ mixing terms.

The eigenvalue equations for the unperturbed $(\eta = 0)$ mass squared operator $\mathcal{O}_0$, acting on the right-handed sector, are

$$-2\nabla_-\nabla_+\psi_{+R} = M^2\psi_{+R},$$
$$-2\nabla_+\nabla_-\psi_{-R} = M^2\psi_{-R}, \tag{B.3.1}$$

The differential equation (B.3.1) can be transformed in an algebraic one through the harmonic expansion, remembering the iso-helicities of ψ_{+R} and ψ_{-R}: $\lambda_{+R} = \lambda_{-R} = 1$. Therefore an explicit expression for the vectors $|i>$, which satisfies by definition $\mathcal{O}_0|i>=0$, is given by

$$|i>=\begin{pmatrix}0\\ \sqrt{\frac{3}{4\pi}}\mathcal{D}_{-1,i}^{(1)}\end{pmatrix}, \qquad i = 1, -1, 0. \tag{B.3.2}$$

We give also an expression for the vectors $|\tilde{i}>$ and the corresponding non vanishing eigenvalues $M_{\tilde{i}}^2$. For $l = 1$, we have just one eigenvalue $M^2 = 2/a^2$ and the corresponding eigenvectors are

$$|\tilde{1}_m>=\begin{pmatrix}\sqrt{\frac{3}{4\pi}}\mathcal{D}_{-1,m}^{(1)}\\ 0\end{pmatrix}. \tag{B.3.3}$$

---

[4]In order to see that we use the background constraints (3.2.4).

For $l = 2$ we have an eigenvalue $M^2 = 6/a^2$, which corresponds to the eigenvectors

$$|\tilde{2}_m> = \begin{pmatrix} \sqrt{\frac{5}{4\pi}} \mathcal{D}^{(2)}_{-1,m} \\ 0 \end{pmatrix}, \tag{B.3.4}$$

and an eigenvalue $M^2 = 4/a^2$, which corresponds to the eigenvectors

$$|\tilde{3}_m> = \begin{pmatrix} 0 \\ \sqrt{\frac{5}{4\pi}} \mathcal{D}^{(1)}_{-1,m} \end{pmatrix}. \tag{B.3.5}$$

By inserting these eigenvectors and eigenvalues in the expression (3.3.39) we get

$$M^2_{ij} = diag\left(0, \ 0, \ \frac{2}{3}|\eta|g_Y^2 \frac{\beta}{\kappa^2} \right), \tag{B.3.6}$$

which corresponds to the spectrum we discussed at the end of section 3.3.3.

Appendix C

Spectrum from 6D Gauged Minimal Supergravity

C.1 Stability Analysis for Sphere Compactification

Here we consider the present known anomaly free 6D gauged minimal supergravities that we discussed in Subsection 4.2.2. The bosonic action for this class of supergravities is given in (4.2.30) and we restrict to the case in which $\phi^\alpha = 0$. We expect that this set up supports stability as we discussed in Subsection 4.2.2. In this section we study when the monopole $(Minkowski)_4 \times S^2$ compactification, given in (4.2.32), is stable, in the sense that there are no tachyons. The only cases in which we can have stability are the old $E_7 \times E_6 \times U(1)_R$ model and a new $SU(2) \times U(1)_R$ model with a particular hyperinos representation (up to the trivial case in which the monopole is embedded in a $U(1)$ factor of the gauge group). So a stable compactification of this type and the embedding of the SM gauge group is possible only in the $E_7 \times E_6 \times U(1)_R$ case.

C.1.1 The Light Cone Gauge

A first step toward the stability analysis is deriving the lagrangian for the small fluctuations around the background. Here we give explicitly such lagrangian for the smooth sphere compactification and its form in the light cone gauge[1]. We do not include hyperscalars in our analysis as they do not mix with the rest and they cannot contain tachyons: this a consequence of the fact that the 6D potential has a global minimum in $\phi^\alpha = 0$ [60, 65] and the Laplacian on the internal space gives a positive contribution to the squared hyperscalar masses.

We denote by h_{MN}, V_M and σ' the metric, gauge field and dilaton fluctuations[2] around the background. The expression of G_3 at the linear level in the fluctuations is

$$G_3 = dV_2 + 2\bar{F} \wedge V, \tag{C.1.1}$$

where the 2-form fluctuation V_2, whose components are V_{MN}, is defined by

$$V_2 \equiv \kappa \left(B_2 - \bar{A} \wedge V \right) \tag{C.1.2}$$

and $\bar{F}$ is the field strength of $\bar{A}$ defined in (4.2.32). The bilinear bosonic action for the fluctuations h_{MN}, V_M, V_{MN} and σ' around the background (4.2.32) reads

$$
\begin{aligned}
S_{B2} = \int d^6 X \sqrt{-G} \Big\{ &\frac{1}{4} h_{MN} \nabla^2 h^{MN} - \frac{1}{8} h \nabla^2 h \\
&+ \frac{1}{2} \nabla^N \left(h_{MN} - \frac{1}{2} G_{MN} h \right) \nabla_R \left(h^{MR} - \frac{1}{2} G^{MR} h \right) + \frac{1}{2} R_{MN} h^{MR} h^N_R \\
&+ \kappa \bar{F}_{MN} \left[-\frac{1}{2} \nabla^M V^N (h + \sigma') + \left(\nabla_R V^N - \nabla^N V_R \right) h^{RM} \right] \\
&+ \frac{1}{2} V_M \nabla^2 V^M + \frac{1}{2} R_{MN} V^M V^N + \frac{1}{2} \left(\nabla_M V^M \right)^2 - \bar{g} \bar{F}_{MN} V^M \times V^N \\
&- \frac{1}{48} \left(\nabla_{[M} V_{NR]} \right)^2 - \frac{\kappa}{12} \nabla_{[M} V_{NR]} V^{[M} \bar{F}^{NR]} - \frac{\kappa^2}{12} \left(V_{[M} \bar{F}_{NR]} \right)^2 \\
&+ \frac{1}{4} \sigma' \left(\nabla^2 - \frac{1}{a^2} \right) \sigma' - \frac{1}{2} \sigma' R_{MN} h^{MN} \Big\},
\end{aligned} \tag{C.1.3}
$$

where the indices are raised and lowered by the background metric G_{MN}, moreover $h \equiv h^M_M$ and we have introduced the following notations

$$\nabla_{[M} V_{NR]} \equiv \nabla_M V_{NR} + \nabla_N V_{RM} + \nabla_R V_{MN}, \tag{C.1.4}$$

$$(V_M \times V_N)^I = -f^{JKI} V^J_M V^K_N, \tag{C.1.5}$$

where f^{JKI} are the structure constants[3] of $\mathcal{G}$.

---

[1] For an introduction to the light cone gauge in higher dimensional field theories see [52, 53, 54].

[2] The fluctuations are properly normalized in a way that their kinetic terms are canonical.

[3] We take the generators T^I of $\mathcal{G}$ satisfying $[T^I, T^J] = if^{IJK} T^K$ and $Tr(T^I T^J) = \delta^{IJ}$.

To study the physical spectrum in a proper way we have to remove the gauge freedom of S_{B2} under 6D diffeomorphisms and gauge transformations. This can be achieved by fixing the light cone gauge, which is defined by $h_{-M} = V_- = V_{-M} = 0$, where the $\pm$ components of a vector A^M are $A^{\pm} \equiv \frac{1}{\sqrt{2}}(A^0 \pm A^3)$. The light cone gauge advantage is the sectors with different helicities decouple and one can easily find the physical degrees of freedom. In the light cone gauge the action (C.1.3) becomes

$$
\begin{aligned}
S_{B2} = \int d^6 X \sqrt{-G} \Bigg\{ & \frac{1}{4} h_{ij}^t \nabla^2 h_{ij}^t + \frac{1}{2} h_{i\alpha} \left(\nabla^2 - \frac{1}{a^2} \right) h_{i\alpha} + \frac{1}{2} V_i \nabla^2 V_i \\
& + \frac{1}{8} V_{i\alpha} \left(\nabla^2 - \frac{1}{a^2} \right) V_{i\alpha} - \kappa \bar{F}_{\alpha\beta} \nabla_\beta V_i h_{i\alpha} - \frac{\kappa}{2} \bar{F}_{\alpha\beta} V_i \nabla_\beta V_{i\alpha} \\
& - \frac{\kappa^2}{4} (V_i \bar{F}_{\alpha\beta})^2 + \frac{1}{4} h_{\alpha\beta} \left(\nabla^2 - \frac{2}{a^2} \right) h_{\alpha\beta} + \frac{1}{8} h_{\alpha\alpha} \nabla^2 h_{\beta\beta} \\
& - \frac{\kappa^2}{2} (\bar{F}_{\alpha\beta} V_\beta)^2 - \kappa \bar{F}_{\alpha\beta} \nabla_\beta V_\gamma h_{\alpha\gamma} - \frac{\kappa}{2} \bar{F}_{\alpha\beta} \nabla_\alpha V_\beta \sigma' \\
& + \frac{1}{2} V_\alpha \left(\nabla^2 - \frac{1}{a^2} \right) V_\alpha - \bar{g} \bar{F}_{\alpha\beta} (V_\alpha \times V_\beta) - \frac{\kappa^2}{32} (V_{\alpha\beta} \bar{F}_{\alpha\beta})^2 \\
& + \frac{1}{16} V_{ij} \nabla^2 V_{ij} + \frac{1}{16} V_{\alpha\beta} \nabla^2 V_{\alpha\beta} - \frac{\kappa}{2} \bar{F}_{\beta\gamma} \nabla_\alpha \nabla_\gamma V_{\alpha\beta} \\
& + \frac{1}{4} \sigma' \left(\nabla^2 - \frac{1}{a^2} \right) \sigma' + \frac{1}{2a^2} \sigma' h_{\alpha\alpha} \Bigg\}, \quad\quad \text{(C.1.6)}
\end{aligned}
$$

where $i, j, k, \dots$ label the transverse 4D coordinates $(i, j, k = 1, 2)$, $\alpha, \beta, \gamma, \dots$ label a local orthonormal basis in the internal space and $h_{ij}^t \equiv h_{ij} - \frac{1}{2} \delta_{ij} h_{kk}$. Apart from the hyperscalars, the complete bosonic 4D spectrum coming from the compactification (4.2.32) can be computed through (C.1.6) by using the harmonic expansion over S^2 [7], that we discussed in Subsection 3.2.1. The main result is the presence of a massless graviton, 4D gauge fields of the group $\mathcal{H}$, which is defined in Subsection 4.2.2, 4D gauge fields of the internal manifold isometries and some scalar fields. Our aim is to study the latter sector which is of course the possible source of tachyons.

C.1.2   Stability Analysis

To study the stability of background (4.2.32) we focus on the helicity-0 terms of the action S_{B2} in the light cone gauge:

$$
\begin{aligned}
S_0 \; = \; & \int d^6 X \sqrt{-G} \left\{ \frac{1}{4} h_{\alpha\beta} \left(\nabla^2 - \frac{2}{a^2} \right) h_{\alpha\beta} + \frac{1}{8} h_{\alpha\alpha} \nabla^2 h_{\beta\beta} \right. \\
& - \frac{\kappa^2}{2} \left(\bar{F}_{\alpha\beta} V_\beta \right)^2 - \kappa \bar{F}_{\alpha\beta} \nabla_\beta V_\gamma h_{\alpha\gamma} - \frac{\kappa}{2} \bar{F}_{\alpha\beta} \nabla_\alpha V_\beta \sigma' \\
& + \frac{1}{2} V_\alpha \left(\nabla^2 - \frac{1}{a^2} \right) V_\alpha - \bar{g} \bar{F}_{\alpha\beta} \left(V_\alpha \times V_\beta \right) - \frac{\kappa^2}{32} \left(V_{\alpha\beta} \bar{F}_{\alpha\beta} \right)^2 \\
& + \frac{1}{16} V_{\alpha\beta} \nabla^2 V_{\alpha\beta} - \frac{\kappa}{2} \bar{F}_{\beta\gamma} V_\alpha \nabla_\gamma V_{\alpha\beta} + \frac{1}{4} \sigma' \left(\nabla^2 - \frac{1}{a^2} \right) \sigma' \\
& \left. + \frac{1}{16} V_{ij} \nabla^2 V_{ij} + \frac{1}{2a^2} \sigma' h_{\alpha\alpha} \right\} .
\end{aligned}
\tag{C.1.7}
$$

The action (C.1.7) contains all the helicity-0 fields, including the helicity-0 components of spin-1 and spin-2 objects. In particular (C.1.7) includes all the physical scalar fields which a priori could be tachyonic.

We observe that the field V_{ij} do not contain tachyons because its spectrum is simply $a^2 M^2 = l(l+1)$, $l = 0, 1, 2, 3$. The scalars V_α, coming from the 6D gauge field fluctuations, can be decomposed in the following three pieces

$$
V_\alpha = \left(V_\alpha', V_\alpha^0, v_\alpha \right)
\tag{C.1.8}
$$

where V_α' is along the generators of $\mathcal{H}$, V_α^0 is along the monopole and v_α is along the generators which do not commute with Q. The simplest sector is V_α' since it does not mix with other fields as it is clear from (C.1.7). The bilinear action for these fields is simply $\frac{1}{2} V_\alpha' \left(\nabla^2 - 1/a^2 \right) V_\alpha'$ and the squared masses are $a^2 M^2 = l(l+1)$, with $l = 1, 2, 3, ...$; therefore here we do not have tachyons. The sector including V_α^0 is much more complicated as this field mixes with other degrees of freedom: the complete sector is given by

$$
\left(V_\alpha^0, h_{\alpha\beta}, V_{\alpha\beta}, \sigma' \right),
\tag{C.1.9}
$$

in the sense that the fields given in (C.1.9) mix each other but do not mix with additional helicity-0 fields, as it can be easily deduced from (C.1.7). So in general we have a mixing between the dilaton, the scalars coming from the 6D gauge field along the monopole and from the 2-form B_2 and the *graviscalars*. The complete spectrum of this sector is given in Table C.1. We observe that there are no tachyons but we have a massless scalar field which corresponds to the first row of Table C.1.

To study the stability properties of solution (4.2.32) the most interesting sector is v_α because in general it can contain tachyons [67]. Let us summarize the result of [67] as it will be useful for our analysis. From formula (C.1.7)

$a^2 M^2$	Multiplicity	range of l
0	1	$l = 0$
2	2	$l = 0$
2	3	$l = 1$
6	2	$l = 1$
$l(l+1)$	3	$l \geq 2$
$l(l-1)$	2	$l \geq 2$
$l(l+1) + 2(l+1)$	2	$l \geq 2$

Table C.1: The spectrum of the $(V_\alpha^0, h_{\alpha\beta}, V_{\alpha\beta}, \sigma')$ sector, including all the helicity-0 fields. The multiplicity in the second column is given in unit of $2l+1$ where l is a non negative integer.

one can easily understand that v_α does not mix with the rest and its bilinear action is

$$\int d^6 X \sqrt{-G} \left\{ \frac{1}{2} v_\alpha \left(\nabla^2 - \frac{1}{a^2} \right) v_\alpha - \bar{g} \bar{F}_{\alpha\beta} (v_\alpha \times v_\beta) \right\}. \qquad (\text{C.1.10})$$

We denote by T^i the generators of $\mathcal{G}$ which do not commute with Q and we choose T^i to be a basis of eigenvectors of the adjoint representation of Q:

$$[Q, T^i] = q_i T^i. \qquad (\text{C.1.11})$$

The qs represent the charges of v_α under[4] $U(1)_M$. By using again the S^2 harmonic expansion one finds the following KK tower

$$a^2 M^2 = l(l+1) - \left(\frac{nq}{2} \right)^2, \qquad (\text{C.1.12})$$

where $l = |1 \pm |nq/2|| + k$, with $k = 0, 1, 2, 3, \dots$ Therefore we have tachyons whenever $|nq| > 1$. We observe that, for a given model, the product nq has to be an integer for each representation because of the Dirac quantization condition. This implies that we have tachyons whenever $|q|$ assumes more than one value.

Now we want to apply these results to all the present anomaly free models of this type [59, 67, 68, 70]. We summarize their structure as follows.

I The first anomaly free model was given in [59] and it has $\mathcal{G} = E_7 \times E_6 \times U(1)_R$ and $R_H = (\mathbf{912}, \mathbf{1})$, where R_H is the hyperinos representation.

II Another example is given in [67] where $\mathcal{G} = E_7 \times G_2 \times U(1)_R$ and $R_H = (\mathbf{56}, \mathbf{14})$.

III In Ref. [68] we have $\mathcal{G} = F_4 \times Sp(9) \times U(1)_R$ and $R_H = (\mathbf{52}, \mathbf{18})$.

---

[4]We remind that $U(1)_M$ is the Abelian group in the direction of the monopole.

IV Finally in Ref. [70] a huge number of anomaly free models was found with $\mathcal{G}$ given by products of $U(1)$ and/or $SU(2)$ and particular hyperinos representations.

We observe that models I,II and III have an enough large $\mathcal{G}$ to include the standard model gauge group, whereas the models IV have not. However, the results of [70] prove that $\mathcal{G}$ do not have to include exceptional groups from the pure mathematical point of view.

The stability analisis for models I and II was already done in [59, 67, 68] so we briefly summarize the result. If we consider model I, the only stable embedding of Q in a non-Abelian algebra[5] is $Q \subset Lie(E_6)$. Whereas, in model II, we have no non-Abelian stable embeddings.

The stability analysis for the remaining models III and IV was never performed. So here we study these cases in detail. Let us consider first model III. In this case the non-abelian embeddings can be $Q \subset Lie(F_4)$ or $Q \subset Lie(Sp(9))$. In the case $Q \subset Lie(F_4)$ we have instability because the system of roots of F_4, which is nothing but all the possible value of q by definition of roots, is given by

$$(\pm 1, \pm 1, 0, 0)$$
$$(\pm 1, 0, \pm 1, 0)$$
$$(\pm 1, 0, 0, \pm 1)$$
$$(0, \pm 1, \pm 1, 0)$$
$$(0, \pm 1, 0, \pm 1)$$
$$(0, 0, \pm 1, \pm 1)$$
$$(\pm 1, 0, 0, 0)$$
$$(0, \pm 1, 0, 0)$$
$$(0, 0, \pm 1, 0)$$
$$(0, 0, 0, \pm 1)$$
$$(\pm \frac{1}{2}, \pm \frac{1}{2}, \pm \frac{1}{2}, \pm \frac{1}{2}). \tag{C.1.13}$$

Each column of system (C.1.13) corresponds to a choice of Q in the Cartan subalgebra of F_4, which is indeed 4-dimensional. Each row of (C.1.13) correspond to a value of q. We note that for every embedding of Q in the Cartan subalgebra of F_4 we have at least the values $q = 1$ and $q = 1/2$, which is enough to conclude that the embedding $Q \subset Lie(F_4)$ is unstable.

We consider now $Q \subset Lie(Sp(9))$ and we prove that this embedding is

---

[5]The embedding of Q in an abelian algebra, that is $Q = Lie(U(1))$, is obviously stable because all the q_i vanish.

unstable as well. The Lie algebra[6] of $Sp(n)$ is generated by

$$I_2 \times A, \quad \sigma_1 \times S_1, \quad \sigma_2 \times S_2, \quad \sigma_3 \times S_3, \tag{C.1.14}$$

where I_2 is the 2×2 identity matrix, σ_i the Pauli matrices, A a generic antisymmetric $n \times n$ matrix, and S_i are generic symmetric $n \times n$ matrices. We observe that $I_2 \times A$ and $\sigma_3 \times S_3$ generate an $SU(n) \times U(1)$ subalgebra and we can take, without loss of generality, the Cartan subalgebra of $Sp(n)$ equal to the Cartan subalgebra of $SU(n) \times U(1)$. Of course we have several possibilities to embed Q in such Cartan subalgebra. We consider first $Q = Lie(U(1))$. In this case we observe that the representation **18**, which appears in the hyperinos representations, is the fundamental of $Sp(9)$ and we have the following tensor product

$$\mathbf{2n} \times \mathbf{2n} = \mathbf{Adj} + \mathbf{D^2} + \mathbf{1}, \tag{C.1.15}$$

where **Adj** is the adjoin representation of $Sp(n)$ and $\mathbf{D^2}$ and **1** are irriducible representations coming from the antisymmetric part of $\mathbf{2n} \times \mathbf{2n}$. On the other hand we have the following branching rules with respect to $Sp(n) \to SU(n) \times U(1)$

$$\mathbf{2n} \to \mathbf{n_1} + \bar{\mathbf{n}}_{-1} \tag{C.1.16}$$

and

$$\mathbf{Adj} \to \mathbf{Adj}_{SU(n)} + \mathbf{R}, \tag{C.1.17}$$

where **1** and -1 in (C.1.16) represent the values of q for the representation **2n** in a particular normalization[7] and **R** in (C.1.17) is some representation of $SU(n)$ which can be reducible or irreducible. By putting (C.1.16) and (C.1.17) in (C.1.15) we get

$$\mathbf{R} + \mathbf{D^2} = \mathbf{n_1} \times \mathbf{n_1} + \bar{\mathbf{n}}_{-1} \times \bar{\mathbf{n}}_{-1} + \bar{\mathbf{n}}_{-1} \times \mathbf{n_1}. \tag{C.1.18}$$

Since

$$dim(\mathbf{D^2}) < dim(\mathbf{n_1} \times \mathbf{n_1} + \bar{\mathbf{n}}_{-1} \times \bar{\mathbf{n}}_{-1}), \tag{C.1.19}$$

necessarily **R** contains some representation with charge $|q| = 2$ and therefore this embedding is unstable. The case in which $Q \subset Lie(SU(n))$ can be studied in a similar way and this embedding turns out to be unstable as well.

Finally we consider the models IV and we focus on the case in which $\mathcal{G}$ contains a non-Abelian $SU(2)$ subgroup. All the hyperinos representations of models IV belong to a $(2l + 1)$-dimensional representations of $SU(2)$. By

---

[6]For a more complete discussion on the Lie algebra of $Sp(n)$ see for example [102].

[7]Of course the normalization of the generators is conventional and it cannot change the final result.

using a similar argument we find a stable $Q \subset SU(2)$ embedding only if we have no even $(2l+1)$-dimensional representation; this is a consequence of the fact that even representations correspond to half-integer spin, whereas the adjoint representation of $SU(2)$ has spin 1. In Ref. [70] one model which satisfies this property is given and it has $\mathcal{G} = SU(2) \times U(1)_R$ and the following hyperinos representations: 7 representations **3**, 2 representations **5** and 31 representations **7**.

C.2   Delta-Function Singularities

In this appendix we briefly review how the Ricci scalar acquires a delta-function contribution in the presence of a deficit angle, and examine what this implies for our choice of metric function e^B in (5.1.1). We are going to use some results presented in Section 1.6.

Let us consider the ansatz (5.1.1,5.1.2), and illustrate the case for the deficit angle δ at $\rho = 0$. Near $\rho = 0$ the metric ds_2^2 of the 2D internal space can be written as follows

$$ds_2^2 = d\rho^2 + \left(1 - \frac{\delta}{2\pi}\right)^2 \rho^2 d\varphi^2 . \tag{C.2.1}$$

By using the change of coordinate $r^{1-\delta/2\pi}/(1 - \delta/2\pi) = \rho$, this metric becomes

$$ds^2 = r^{-\delta/\pi} \left(dr^2 + r^2 d\varphi^2\right) . \tag{C.2.2}$$

From (C.2.2) and (1.6.59) one can show

$$R = 2\,\delta\, r^{\delta/\pi} \delta^{(2)}\left(y\right) + ..., \tag{C.2.3}$$

where the 2D vector y is defined by $y = (r\cos\varphi, r\sin\varphi)$, $\delta^{(2)}$ is the 2D Dirac delta-function and the dots are the smooth contributions[8]. On the other hand, near to $\rho = 0$ the Ricci scalar, R, can be expressed in terms of derivatives of B:

$$R = -B'' - \frac{1}{2}(B')^2, \tag{C.2.4}$$

where $' \equiv \partial_\rho$, and from (C.2.3) and (C.2.4) it follows that

$$B'' = -2\,\delta\, r^{\delta/\pi} \delta^{(2)}\left(y\right) + ..., \tag{C.2.5}$$

That is, the metric function e^B contains a delta-function contribution in its

---

[8] Eq. (C.2.3) describes the asymptotic behaviour in the vicinity of the brane [33]. It seems that for positive δ the Ricci scalar vanishes at the origin. However, the first term on the right hand side of (B.3) should be interpreted in a distributional sense. The effect of the Ricci scalar as a distribution on a scalar test function $f(y)$ is then $\int d^2y \sqrt{\bar{g}} R f(y) = 2\,\delta f(0) + ...$

second order derivative with respect to ρ. In coordinate system (4.4.40), (C.2.5) becomes

$$\partial_u^2 B = -2\,\delta\,r^{\delta/\pi}\delta^{(2)}\,(y) + ..., \qquad (C.2.6)$$

The delta-function in the curvature also gives rise to a delta-function in the derivative of the spin connection (5.3.6). The Riemann tensor is defined in terms of the spin connection as:

$$R^A_{\ B} = d\omega^A_{\ B} + \omega^A_{\ C} \wedge \omega^C_{\ B}\,. \qquad (C.2.7)$$

Near to the brane $\rho = 0$, relation (C.2.7) gives $R^5_{\rho\varphi\ 6} = \partial_\rho \omega^5_{\varphi\ 6}$, and from the expression for the spin-connection (5.3.6):

$$\Omega' = -\left(\frac{1}{2}B'' + \frac{1}{4}(B')^2\right)e^{B/2}, \qquad (C.2.8)$$

thus leading to the delta-function behaviour from (C.2.5).

These results must be recalled when obtaining the Schroedinger-like equations that govern the fluctuations.

C.3 Imposing Boundary Conditions

Here we study the implications of the NC and the HC for gauge field fluctuations. To impose the boundary conditions the following properties will be useful:

$$F(a, b, c, z) \overset{z\to 0}{\to} 1, \qquad (C.3.1)$$

$$F(a, b, c, z) = \Gamma_1 F(a, b, a + b - c + 1, 1 - z)$$
$$+\Gamma_2(1 - z)^{c-a-b} F(c - a, c - b, c - a - b + 1, 1 - z), \qquad (C.3.2)$$

where

$$\Gamma_1 \equiv \frac{\Gamma(c)\Gamma(c - a - b)}{\Gamma(c - a)\Gamma(c - b)}, \quad \Gamma_2 \equiv \frac{\Gamma(c)\Gamma(-c + a + b)}{\Gamma(a)\Gamma(b)}, \qquad (C.3.3)$$

and Γ is the Euler gamma function. The relation (C.3.2) is valid if $c - a - b$ is not an integer [103] and y_1 and y_2 in (5.2.21) are both well defined when c is not an integer. In general $c - a - b$ and c are not integers for generic ω and $\bar\omega$; so we can consider ω and $\bar\omega$ as regulators to use (5.2.21) and (C.3.2) and at the end we can take the limits in which $c - a - b$ and c go to an integer, which will turn out to be well defined.

We first consider the behaviour of ψ for $u \to \bar u$, that is $z \to 0$ because of the definition $z = \cos^2\left(\frac{u}{r_0}\right)$. For $c \neq 1$ we use the expression for ψ given in (5.2.22) and property (C.3.1) gives us

$$\psi \overset{u \to \overline{u}}{\to} K_1(\overline{u} - u)^{2\gamma} + K_2(\overline{u} - u)^{1-2\gamma}, \tag{C.3.4}$$

where we used $c = 1/2 + 2\gamma$. So the NC (5.2.5) implies $K_1 = 0$ when $\gamma \leq -1/4$ and $K_2 = 0$ when $\gamma \geq 3/4$. On the other hand the HC (5.2.16) implies[9]

$$\lim_{u \to \overline{u}} \psi^* \left(-\partial_u + \frac{1}{2} \frac{1}{u - \overline{u}} \right) \psi < \infty \tag{C.3.5}$$

and by using the behaviour (C.3.4) this limit becomes

$$\left(2\gamma - \frac{1}{2} \right) \lim_{u \to \overline{u}} \left[|K_1|^2 (\overline{u} - u)^{4\gamma - 1} - K_1^* K_2 + K_1 K_2^* - |K_2|^2 (\overline{u} - u)^{-4\gamma + 1} \right], \tag{C.3.6}$$

so the HC implies $K_1 = 0$ when $\gamma < 1/4$ and $K_2 = 0$ when $\gamma > 1/4$. The case $\gamma = 1/4$ corresponds to $c = 1$ and so we have to use the expression of ψ given in (5.2.24). We have then

$$\psi \overset{u \to \overline{u}}{\to} K_1(\overline{u} - u)^{1/2} - K_2(\overline{u} - u)^{1/2} \ln(\overline{u} - u) \tag{C.3.7}$$

for which (C.3.5) implies $K_2 = 0$. Therefore we obtain (5.2.25) and (5.2.26).

The discreteness of the spectrum emerges when we impose the NC and HC for $u \to 0$. For instance for $m \geq N_V$, up to an overall constant, the behaviour of ψ is given by properties (C.3.1) and (C.3.2):

$$\psi \overset{u \to 0}{\to} \Gamma_1 u^{2\beta} + \Gamma_2 u^{1-2\beta}, \tag{C.3.8}$$

where $\Gamma_{1,2}$ are defined in (C.3.3) and we used $c - a - b = 1/2 - 2\beta$. Behaviour (C.3.8) is similar to (C.3.4) but γ is replaced by β. So, following the same steps as above, the NC and the HC imply that $\Gamma_1 = 0$ for $\beta < 1/4$ and $\Gamma_2 = 0$ for $\beta > 1/4$. Let us study the case $m \geq N_V$ and $\beta < 1/4$, that is

$$N_V \leq m < 0. \tag{C.3.9}$$

We then have

$$0 = \Gamma_1 \equiv \frac{\Gamma(c)\Gamma(c - a - b)}{\Gamma(c - a)\Gamma(c - b)}. \tag{C.3.10}$$

Since the Euler gamma function never vanishes we require that $\Gamma(c-a) = \infty$ or $\Gamma(c - b) = \infty$ and this is possible only when $c - a = -n$ or $c - b = -n$, where $n = 0, 1, 2, 3, ...$ By using the definitions (5.2.19) both conditions

---

[9]In fact, for the gauge field sector, each term in the HC (5.2.16) is separately zero if one requires their finiteness. Therefore (C.3.5) and its counter-part for $u = 0$ are sufficient to ensure (5.2.16). On the other hand, for the fermions, there are some cases in which they are each finite and non-zero, and so (5.2.16) requires that they cancel.

lead to the following squared masses

$$M_{V\,n,m}^2 = \frac{4}{r_0^2}\left\{n(n+1) + \left(\frac{1}{2}+n\right)[-m\omega + (m-N_V)\overline{\omega}]\right\} \quad \text{(C.3.11)}$$

which are positive because of (C.3.9) and $n \geq 0$. When $m \geq N_V$ and[10] $\beta \geq 1/4$, that is

$$m \geq N_V \quad \text{and} \quad m \geq 0, \quad \text{(C.3.12)}$$

we have

$$0 = \Gamma_2 \equiv \frac{\Gamma(c)\Gamma(-c+a+b)}{\Gamma(a)\Gamma(b)} \quad \text{(C.3.13)}$$

and this implies $a = -n$ or $b = -n$, where $n = 0, 1, 2, 3, \dots$. The corresponding squared masses are

$$M_{V\,n,m}^2 = \frac{4}{r_0^2}\left\{n(n+1) + \left(\frac{1}{2}+n\right)[m\omega + (m-N_V)\overline{\omega}] + m\omega(m-N_V)\overline{\omega}\right\}$$
$$\text{(C.3.14)}$$

which are positive or vanishing. We can study the case $m < N_V$ in a similar way. The complete result for the gauge fields sector is given in equations (5.2.27)-(5.2.30). We observe that (5.2.16) is now automatically satisfied by every pair of wave functions ψ and ψ', for a given quantum number m, since the asymptotic behaviour of the wave function cannot depend on the quantum number n: this is a consequence of the $1/u^2$ and $1/(\overline{u}-u)^2$ singularities of the potential V in (5.2.14).

C.4   Complete Fermionic Mass Spectrum

In this appendix we give the complete fermionic spectrum which is also labeled by an integer quantum number $n = 0, 1, 2, \dots$. Although much longer, the calculation proceeds in exactly the same way as for the gauge field sector, outlined in the previous appendix[11].

$m \geq -\frac{1}{2} + N + \frac{1}{2\overline{\omega}}$: in this case $K_2 = 0$ and we get the following squared masses.

---

[10]The $\beta = 1/4$ case is recovered by taking the limit $\omega \to 0$.

[11]There is one additional subtlety. Here, for the values of m which allow a zero mode, we must impose a mixed HC between the massless mode and massive modes, in addition to the diagonal HC. In general the HC involving distinct wave functions ψ_{mn} and $\psi_{mn'}$ does not lead to additional constraints, because the asymptotic behaviour of the modes is independent of n. However, the massless modes are more strongly constrained than the massive ones, obeying as they do a decoupled Dirac equation in addition to the Schroedinger equation.

- For $m > -\frac{1}{2} + \frac{1}{2\omega}$

$$M^2_{F\,n,m} = \frac{4}{r_0^2}\left[\frac{1}{2} + n + \omega\left(\frac{1}{2} + m\right)\right]$$
$$\times \left[\frac{1}{2} + n + \overline{\omega}\left(\frac{1}{2} + m - N\right)\right] > 0. \qquad (C.4.1)$$

- For $-\frac{1}{2} - \frac{1}{2\omega} < m < -\frac{1}{2} + \frac{1}{2\omega}$

$$M^2_{F\,n,m} = \frac{4}{r_0^2}\left[\frac{1}{2} + n + \omega\left(\frac{1}{2} + m\right)\right]$$
$$\times \left[\frac{1}{2} + n + \overline{\omega}\left(\frac{1}{2} + m - N\right)\right] > 0. \qquad (C.4.2)$$

  or

$$M^2_{F\,n,m} = \frac{4}{r_0^2}(1+n)\left[1 + n + \overline{\omega}\left(\frac{1}{2} + m - N\right)\right.$$
$$\left. - \omega\left(\frac{1}{2} + m\right)\right] > 0. \qquad (C.4.3)$$

- For $m \leq -\frac{1}{2} - \frac{1}{2\omega}$

$$M^2_{F\,n,m} = \frac{4}{r_0^2}(1+n)\left[1 + n + \overline{\omega}\left(\frac{1}{2} + m - N\right)\right.$$
$$\left. - \omega\left(\frac{1}{2} + m\right)\right] > 0. \qquad (C.4.4)$$

$m \leq -\frac{1}{2} + N - \frac{1}{2\overline{\omega}}$: in this case $K_1 = 0$ and we get the following squared masses.

- For $m > -\frac{1}{2} + \frac{1}{2\omega}$

$$M^2_{F\,n,m} = \frac{4}{r_0^2}n\left[n - \overline{\omega}\left(\frac{1}{2} + m - N\right) + \omega\left(\frac{1}{2} + m\right)\right] \geq 0.$$
$$(C.4.5)$$

- For $-\frac{1}{2} - \frac{1}{2\omega} < m < -\frac{1}{2} + \frac{1}{2\omega}$

$$M^2_{F\,n,m} = \frac{4}{r_0^2}n\left[n - \overline{\omega}\left(\frac{1}{2} + m - N\right) + \omega\left(\frac{1}{2} + m\right)\right] \geq 0.$$
$$(C.4.6)$$

- For $m \leq -\frac{1}{2} - \frac{1}{2\omega}$

$$M_{F\,n,m}^2 = \frac{4}{r_0^2}\left[\frac{1}{2} + n - \omega\left(\frac{1}{2} + m\right)\right]$$
$$\times \left[\frac{1}{2} + n - \overline{\omega}\left(\frac{1}{2} + m - N\right)\right] > 0. \qquad (C.4.7)$$

$-\frac{1}{2} + N - \frac{1}{2\overline{\omega}} < m < -\frac{1}{2} + N + \frac{1}{2\overline{\omega}}$: this case is possible only when $\overline{\delta} < 0$.

- For $m > -\frac{1}{2} + \frac{1}{2\omega}$ we have $K_1 = 0$ and

$$M_{F\,n,m}^2 = \frac{4}{r_0^2}n\left[n - \overline{\omega}\left(\frac{1}{2} + m - N\right) + \omega\left(\frac{1}{2} + m\right)\right] \geq 0. \qquad (C.4.8)$$

- For $m \leq -\frac{1}{2} - \frac{1}{2\omega}$ we have two possibilities. We have $K_2 = 0$ and

$$M_{F\,n,m}^2 = \frac{4}{r_0^2}(1 + n)\left[1 + n + \overline{\omega}\left(\frac{1}{2} + m - N\right)\right.$$
$$\left. - \omega\left(\frac{1}{2} + m\right)\right] > 0 \qquad (C.4.9)$$

or $K_1 = 0$ and

$$M_{F\,n,m}^2 = \frac{4}{r_0^2}\left[\frac{1}{2} + n - \omega\left(\frac{1}{2} + m\right)\right]$$
$$\times \left[\frac{1}{2} + n - \overline{\omega}\left(\frac{1}{2} + m - N\right)\right] > 0 \qquad (C.4.10)$$

- $-\frac{1}{2} - \frac{1}{2\omega} < m < -\frac{1}{2} + \frac{1}{2\omega}$: this case is possible only when $\delta < 0$ and we get $K_1 = 0$ and

$$M_{F\,n,m}^2 = \frac{4}{r_0^2}n\left[n - \overline{\omega}\left(\frac{1}{2} + m - N\right) + \omega\left(\frac{1}{2} + m\right)\right] \geq 0. \qquad (C.4.11)$$

Again, we can perform a check of our results by considering the S^2 limit $(\omega \to 1, \overline{\omega} \to 1)$. In this case, the mass spectrum (C.4.1)-(C.4.11) reduces correctly to

$$a^2 M_F^2 = \left(l + \frac{1 + N}{2}\right)\left(l + \frac{1 - N}{2}\right), \quad multiplicity = 2l + 1 \qquad (C.4.12)$$

where $a = r_0/2$ is the radius of S^2 and[12] $l = \frac{|N| - 1}{2} + k$ and $k = 0, 1, 2, 3, \ldots$.

---

[12]The number l is defined in different ways in equations (C.4.1)-(C.4.11). For instance we have $l \equiv n + m + (1 - N)/2$ for (C.4.1) and $l \equiv 1/2 + n - N/2$ for

Bibliography

[1] T. Kaluza, "On The Problem Of Unity In Physics," Sitzungsber. Preuss. Akad. Wiss. Berlin (Math. Phys.) **1921** (1921) 966.

[2] O. Klein, "Quantum Theory And Five-Dimensional Theory Of Relativity," Z. Phys. **37** (1926) 895 [Surveys High Energ. Phys. **5** (1986) 241].

[3] T. Appelquist, A. Chodos and P. G. O. Freund, "Modern Kaluza-Klein Theories," (Frontiers in Physics, Vol 65).

[4] S. Randjbar-Daemi, A. Salvio and M. Shaposhnikov, "On the decoupling of heavy modes in Kaluza-Klein theories," Nucl. Phys. B **741** (2006) 236 [arXiv:hep-th/0601066].
A. Salvio, "4D effective theory and geometrical approach," talk given at 2nd Cairo International Conference on High Energy Physics (CICHEP 2), Cairo, Egypt, 14-17 Jan 2006 [arXiv:hep-th/0609050].

[5] A. Salam and J. A. Strathdee, "On Kaluza-Klein Theory," Annals Phys. **141** (1982) 316.

[6] S. Randjbar-Daemi and R. Percacci, "Spontaneous Compactification Of A (4+D)-Dimensional Kaluza-Klein Theory Into M(4) X G/H For Arbitrary G And H," Phys. Lett. B **117** (1982) 41.

[7] S. Randjbar-Daemi, A. Salam and J. A. Strathdee, "Spontaneous Compactification In Six-Dimensional Einstein-Maxwell Theory," Nucl. Phys. B **214** (1983) 491.

[8] V. A. Rubakov and M. E. Shaposhnikov, "Do We Live Inside A Domain Wall?," Phys. Lett. B **125** (1983) 136.

[9] K. Akama, "An Early Proposal Of 'Brane World'," Lect. Notes Phys. **176** (1982) 267 [arXiv:hep-th/0001113].

[10] H. Nicolai and C. Wetterich, "On The Spectrum Of Kaluza-Klein Theories With Noncompact Internal Spaces," Phys. Lett. B **150** (1985) 347.

[11] G. W. Gibbons and D. L. Wiltshire, "Space-Time As A Membrane In Higher Dimensions," Nucl. Phys. B **287**, 717 (1987) [arXiv:hep-th/0109093].

[12] A. Kehagias, "On non-compact compactifications with brane worlds," arXiv:hep-th/9911134.

[13] R. F. Dashen, B. Hasslacher and A. Neveu, "Nonperturbative Methods And Extended Hadron Models In Field Theory. 2. Two-Dimensional Models And Extended Hadrons," Phys. Rev. D **10** (1974) 4130.

[14] P. M. Morse and H. Feshbach, "Methods of Theoretical Physics," McGraw-Hill, New York, 1953.

[15] R. Rajaraman, "Solitons and Instantons," Amsterdam, North-Holland, 1987.

[16] C. Wetterich, "Chiral Fermions In Six-Dimensional Gravity," Nucl. Phys. B **253** (1985) 366.

[17] S. Randjbar-Daemi and M. E. Shaposhnikov, "Fermion zero-modes on brane-worlds," Phys. Lett. B **492** (2000) 361 [arXiv:hep-th/0008079].

[18] G. R. Dvali and M. A. Shifman, "Domain walls in strongly coupled theories," Phys. Lett. B **396** (1997) 64 [Erratum-ibid. B **407** (1997) 452] [arXiv:hep-th/9612128].

[19] N. Arkani-Hamed, S. Dimopoulos and G. R. Dvali, "The hierarchy problem and new dimensions at a millimeter," Phys. Lett. B **429** (1998) 263 [arXiv:hep-ph/9803315].
I. Antoniadis, N. Arkani-Hamed, S. Dimopoulos and G. R. Dvali, "New dimensions at a millimeter to a Fermi and superstrings at a TeV," Phys. Lett. B **436** (1998) 257 [arXiv:hep-ph/9804398].

[20] E. G. Adelberger, B. R. Heckel and A. E. Nelson, "Tests of the gravitational inverse-square law," Ann. Rev. Nucl. Part. Sci. **53** (2003) 77 [arXiv:hep-ph/0307284].

[21] S. L. Parameswaran, S. Randjbar-Daemi and A. Salvio, "Gauge Fields, Fermions and Mass Gaps in 6D Brane Worlds," accepted for publication in Nucl. Phys. B [arXiv:hep-th/0608074].

[22] G. F. Giudice, R. Rattazzi and J. D. Wells, "Quantum gravity and extra dimensions at high-energy colliders," Nucl. Phys. B **544** (1999) 3 [arXiv:hep-ph/9811291].

[23] L. Randall and R. Sundrum, "A large mass hierarchy from a small extra dimension," Phys. Rev. Lett. **83** (1999) 3370 [arXiv:hep-ph/9905221].

[24] L. Randall and R. Sundrum, "An alternative to compactification," Phys. Rev. Lett. **83** (1999) 4690 [arXiv:hep-th/9906064].

[25] S. Chang, J. Hisano, H. Nakano, N. Okada and M. Yamaguchi, "Bulk standard model in the Randall-Sundrum background," Phys. Rev. D **62** (2000) 084025 [arXiv:hep-ph/9912498].

[26] W. D. Goldberger and M. B. Wise, "Bulk fields in the Randall-Sundrum compactification scenario," Phys. Rev. D **60** (1999) 107505 [arXiv:hep-ph/9907218].
H. Davoudiasl, J. L. Hewett and T. G. Rizzo, "Bulk gauge fields in the Randall-Sundrum model," Phys. Lett. B **473** (2000) 43 [arXiv:hep-ph/9911262].
T. Gherghetta and A. Pomarol, "Bulk fields and supersymmetry in a slice of AdS," Nucl. Phys. B **586** (2000) 141 [arXiv:hep-ph/0003129].

[27] V. A. Rubakov and M. E. Shaposhnikov, "Extra Space-Time Dimensions: Towards A Solution To The Cosmological Constant Problem," Phys. Lett. B **125** (1983) 139.
S. Randjbar-Daemi and C. Wetterich, "Kaluza-Klein Solutions With Noncompact Internal Spaces," Phys. Lett. B **166** (1986) 65.

[28] G. R. Dvali, G. Gabadadze and M. Porrati, "4D gravity on a brane in 5D Minkowski space," Phys. Lett. B **485** (2000) 208 [arXiv:hep-th/0005016].

[29] G. R. Dvali and G. Gabadadze, "Gravity on a brane in infinite-volume extra space," Phys. Rev. D **63** (2001) 065007 [arXiv:hep-th/0008054].

[30] G. Dvali, G. Gabadadze and M. Shifman, "Diluting cosmological constant in infinite volume extra dimensions," Phys. Rev. D **67** (2003) 044020 [arXiv:hep-th/0202174].

[31] R. Sundrum, "Compactification for a three-brane universe," Phys. Rev. D **59** (1999) 085010 [arXiv:hep-ph/9807348].
I. Navarro, "Spheres, deficit angles and the cosmological constant," Class. Quant. Grav. **20** (2003) 3603 [arXiv:hep-th/0305014].

[32] S. M. Carroll and M. M. Guica, "Sidestepping the cosmological constant with football-shaped extra dimensions," arXiv:hep-th/0302067.
I. Navarro, "Codimension two compactifications and the cosmological constant problem," JCAP **0309** (2003) 004 [arXiv:hep-th/0302129].

[33] S. Randjbar-Daemi and V. A. Rubakov, "4d-flat compactifications with brane vorticities," JHEP **0410** (2004) 054 [arXiv:hep-th/0407176].

[34] Y. Aghababaie, C. P. Burgess, S. L. Parameswaran and F. Quevedo, "Towards a naturally small cosmological constant from branes in 6D supergravity," Nucl. Phys. B **680** (2004) 389 [arXiv:hep-th/0304256].

[35] C. P. Burgess, J. Matias and F. Quevedo, "MSLED: A minimal supersymmetric large extra dimensions scenario," Nucl. Phys. B **706** (2005) 71 [arXiv:hep-ph/0404135].

[36] C. P. Burgess and D. Hoover, "UV sensitivity in supersymmetric large extra dimensions: The Ricci-flat case," arXiv:hep-th/0504004.

[37] D. M. Ghilencea, D. Hoover, C. P. Burgess and F. Quevedo, "Casimir energies for 6D supergravities compactified on T(2)/Z(N) with Wilson lines," JHEP **0509** (2005) 050 [arXiv:hep-th/0506164].

[38] D. Hoover and C. P. Burgess, "Ultraviolet sensitivity in higher dimensions," JHEP **0601** (2006) 058 [arXiv:hep-th/0507293].

[39] V. P. Nair and S. Randjbar-Daemi, "Nonsingular 4d-flat branes in six-dimensional supergravities," JHEP **0503** (2005) 049 [arXiv:hep-th/0408063].

[40] J. W. Chen, M. A. Luty and E. Ponton, "A critical cosmological constant from millimeter extra dimensions," JHEP **0009** (2000) 012 [arXiv:hep-th/0003067].

[41] J. C. Pati and A. Salam, "Unified Lepton - Hadron Symmetry, And A Gauge Theory Of The Basic Interactions," Phys. Rev. D **8** (1973) 1240.
H. Georgi and S. L. Glashow, "Unity Of All Elementary Particle Forces," Phys. Rev. Lett. **32** (1974) 438.
H. Georgi, in: "Particle and Fields," (American Institute of Physics, New York, 1975).
H. Fritzsch and P. Minkowski, "Unified Interactions Of Leptons And Hadrons," Annals Phys. **93** (1975) 193.
F. Gursey, P. Ramond and P. Sikivie, "A Universal Gauge Theory Model Based On E6," Phys. Lett. B **60** (1976) 177.
F. Gursey and P. Sikivie, "E(7) As A Universal Gauge Group," Phys. Rev. Lett. **36** (1976) 775.

[42] K. G. Wilson, "The Renormalization Group: Critical Phenomena And The Kondo Problem," Rev. Mod. Phys. **47**, 773 (1975).
S. Weinberg, "Effective Gauge Theories," Phys. Lett. B **91** (1980) 51.

[43] T. Appelquist and J. Carazzone, "Infrared Singularities And Massive Fields," Phys. Rev. D **11** (1975) 2856.

[44] J. C. Collins, F. Wilczek and A. Zee, "Low-Energy Manifestations Of Heavy Particles: Application To The Neutral Current," Phys. Rev. D **18** (1978) 242.

[45] L. H. Chan, T. Hagiwara and B. A. Ovrut, "The Effect Of Heavy Particles In Low-Energy Light Particle Processes," Phys. Rev. D **20** (1979) 1982.

[46] L. F. Li, "Comments On The Decoupling Theorem," Print-80-0554 (CARNEGIE MELLON).

[47] G. W. Gibbons and C. N. Pope, "Consistent S^2 Pauli reduction of six-dimensional chiral gauged Einstein-Maxwell supergravity," Nucl. Phys. B **697** (2004) 225 [arXiv:hep-th/0307052].

[48] A. Salam and E. Sezgin, "Chiral Compactification On $(Minkowski) \times S^2$ Of N=2 Einstein-Maxwell Supergravity In Six-Dimensions," Phys. Lett. B **147** (1984) 47.

[49] A. J. Buras, J. R. Ellis, M. K. Gaillard and D. V. Nanopoulos, "Aspects Of The Grand Unification Of Strong, Weak And Electromagnetic Interactions," Nucl. Phys. B **135** (1978) 66.

[50] E. P. Wigner, "Group Theory and its Application to the Quantum Mechanics of Atomics Spectra," New York: Academic Press, 1959.

[51] J. Sobczyk, "Symmetry Breaking In Kaluza-Klein Theories," Class. Quant. Grav. **4** (1987) 37.

[52] S. Randjbar-Daemi and M. Shaposhnikov, "A formalism to analyze the spectrum of brane world scenarios," Nucl. Phys. B **645** (2002) 188 [arXiv:hep-th/0206016].

[53] S. Randjbar-Daemi, A. Salam and J. A. Strathdee, "Towards A Self-consistent Computation Of Vacuum Energy In Eleven-Dimensional Supergravity," Nuovo Cim. B **84** (1984) 167.

[54] S. Randjbar-Daemi and M. H. Sarmadi, "Graviton Induced Compactification In The Light Cone Gauge," Phys. Lett. B **151** (1985) 343.

[55] C. P. Burgess, "Towards a natural theory of dark energy: Supersymmetric large extra dimensions," AIP Conf. Proc. **743** (2005) 417 [arXiv:hep-th/0411140].
C. P. Burgess, "Supersymmetric large extra dimensions and the cosmological constant problem," arXiv:hep-th/0510123.

[56] A. Salvio, hep-th/0701020.

[57] A. Salvio and M. Shaposhnikov, JHEP **0711** (2007) 037 [arXiv:0707.2455 [hep-th]].

[58] H. Nishino and E. Sezgin, "Matter And Gauge Couplings Of N=2 Supergravity In Six-Dimensions," Phys. Lett. B **144** (1984) 187.

[59] S. Randjbar-Daemi, A. Salam, E. Sezgin and J. A. Strathdee, "An Anomaly Free Model In Six-Dimensions," Phys. Lett. B **151** (1985) 351.

[60] H. Nishino and E. Sezgin, "The Complete N=2, D = 6 Supergravity With Matter And Yang-Mills Couplings," Nucl. Phys. B **278** (1986) 353.

[61] H. Nishino and E. Sezgin, "New couplings of six-dimensional supergravity," Nucl. Phys. B **505** (1997) 497 [arXiv:hep-th/9703075].

[62] Y. Aghababaie, C. P. Burgess, S. L. Parameswaran and F. Quevedo, "SUSY breaking and moduli stabilization from fluxes in gauged 6D supergravity," JHEP **0303** (2003) 032 [arXiv:hep-th/0212091].

[63] G. W. Gibbons, R. Guven and C. N. Pope, "3-branes and uniqueness of the Salam-Sezgin vacuum," Phys. Lett. B **595** (2004) 498 [arXiv:hep-th/0307238].

[64] Y. Aghababaie *et al.*, "Warped brane worlds in six dimensional supergravity," JHEP **0309** (2003) 037 [arXiv:hep-th/0308064].

[65] S. Randjbar-Daemi and E. Sezgin, "Scalar potential and dyonic strings in 6d gauged supergravity," Nucl. Phys. B **692** (2004) 346 [arXiv:hep-th/0402217].

[66] C. P. Burgess, F. Quevedo, G. Tasinato and I. Zavala, "General axisymmetric solutions and self-tuning in 6D chiral gauged supergravity," JHEP **0411** (2004) 069 [arXiv:hep-th/0408109].

[67] S. D. Avramis, A. Kehagias and S. Randjbar-Daemi, "A new anomaly-free gauged supergravity in six dimensions," JHEP **0505** (2005) 057 [arXiv:hep-th/0504033].

[68] S. D. Avramis and A. Kehagias, "A systematic search for anomaly-free supergravities in six dimensions," JHEP **0510** (2005) 052 [arXiv:hep-th/0508172].

[69] H. M. Lee and C. Ludeling, "The general warped solution with conical branes in six-dimensional supergravity," JHEP **0601** (2006) 062 [arXiv:hep-th/0510026].

[70] R. Suzuki and Y. Tachikawa, "More anomaly-free models of six-dimensional gauged supergravity," arXiv:hep-th/0512019.

[71] S. L. Parameswaran, G. Tasinato and I. Zavala, "The 6D SuperSwirl," Nucl. Phys. B **737** (2006) 49 [arXiv:hep-th/0509061].

[72] L. Alvarez-Gaume and E. Witten, "Gravitational Anomalies," Nucl. Phys. B **234** (1984) 269.

[73] A. Salam and E. Sezgin, "Supergravities In Diverse Dimensions. Vol. 1, 2,"

[74] B. de Wit, "Supergravity," arXiv:hep-th/0212245.

[75] A. Van Proeyen, "Structure of supergravity theories," arXiv:hep-th/0301005.

[76] N. Marcus and J. H. Schwarz, "Field Theories That Have No Manifestly Lorentz Invariant Formulation," Phys. Lett. B **115** (1982) 111.

[77] S. Ferrara, F. Riccioni and A. Sagnotti, "Tensor and vector multiplets in six-dimensional supergravity," Nucl. Phys. B **519** (1998) 115 [arXiv:hep-th/9711059].

[78] F. Riccioni, "All couplings of minimal six-dimensional supergravity," Nucl. Phys. B **605** (2001) 245 [arXiv:hep-th/0101074].

[79] C. P. Burgess, "Supersymmetric large extra dimensions and the cosmological constant: An update," Annals Phys. **313** (2004) 283 [arXiv:hep-th/0402200].

[80] A. J. Tolley, C. P. Burgess, D. Hoover and Y. Aghababaie, "Bulk singularities and the effective cosmological constant for higher co-dimension branes," JHEP **0603** (2006) 091 [arXiv:hep-th/0512218].

[81] H. M. Lee and A. Papazoglou, "Scalar mode analysis of the warped Salam-Sezgin model," Nucl. Phys. B **747** (2006) 294 [arXiv:hep-th/0602208].
C. P. Burgess, C. de Rham, D. Hoover, D. Mason and A. J. Tolley, "Kicking the rugby ball: Perturbations of 6D gauged chiral supergravity," arXiv:hep-th/0610078.

[82] M. Giovannini, "Graviphoton and graviscalars delocalization in brane world scenarios," arXiv:hep-th/0111218.

[83] M. Giovannini, J. V. Le Be and S. Riederer, "Zero modes of six-dimensional Abelian vortices," Class. Quant. Grav. **19**, 3357 (2002) [arXiv:hep-th/0205222].

[84] S. Randjbar-Daemi and M. Shaposhnikov, "QED from six-dimensional vortex and gauge anomalies," JHEP **0304**, 016 (2003) [arXiv:hep-th/0303247].

[85] M. F. Manning, "Exact Solutions of the Schroedinger Equation," Phys. Rev. 48 (1935) 161.

[86] J. M. Schwindt and C. Wetterich, "Holographic branes," Phys. Lett. B **578** (2004) 409 [arXiv:hep-th/0309065].

[87] K. R. Dienes, "Shape versus volume: Making large flat extra dimensions invisible," Phys. Rev. Lett. **88** (2002) 011601 [arXiv:hep-ph/0108115].

[88] C. Wetterich, "The Cosmological Constant And Noncompact Internal Spaces In Kaluza-Klein Theories," Nucl. Phys. B **255** (1985) 480.

[89] M. L. Graesser, J. E. Kile and P. Wang, "Gravitational perturbations of a six dimensional self-tuning model," Phys. Rev. D **70**, 024008 (2004) [arXiv:hep-th/0403074].

[90] M. Peloso, L. Sorbo and G. Tasinato, "Standard 4d gravity on a brane in six dimensional flux compactifications," Phys. Rev. D **73** (2006) 104025 [arXiv:hep-th/0603026].

[91] H. Yoshiguchi, S. Mukohyama, Y. Sendouda and S. Kinoshita, "Dynamical stability of six-dimensional warped flux compactification," JCAP **0603**, 018 (2006) [arXiv:hep-th/0512212].

[92] C. de Rham and A. J. Tolley, "Gravitational waves in a codimension two braneworld," JCAP **0602**, 003 (2006) [arXiv:hep-th/0511138].

[93] S. L. Parameswaran, S. Randjbar-Daemi and A. Salvio, JHEP **0801** (2008) 051 [arXiv:0706.1893 [hep-th]].

[94] S. L. Parameswaran, S. Randjbar-Daemi and A. Salvio, JHEP **0903** (2009) 136 [arXiv:0902.0375 [hep-th]].

[95] S. L. Parameswaran, S. Randjbar-Daemi and A. Salvio, arXiv:1001.3271 [hep-th].

[96] A. Salvio, Phys. Lett. B **681** (2009) 166 [arXiv:0909.0023 [hep-th]].

[97] M. Williams, C. P. Burgess, L. van Nierop and A. Salvio, JHEP **1301** (2013) 102 [JHEP **1301** (2013) 102] [arXiv:1210.3753 [hep-th]].

[98] C. P. Burgess, L. van Nierop, S. Parameswaran, A. Salvio and M. Williams, JHEP **1302** (2013) 120 [arXiv:1210.5405 [hep-th]].

[99] A. Salvio, J. Phys. Conf. Ser. **437** (2013) 012004 [arXiv:1210.5852 [hep-th]].

[100] A. N. Schellekens, "Stability Of Higher Dimensional Einstein Yang-Mills Theories On Symmetric Spaces," Nucl. Phys. B **248** (1984) 706.

[101] A. N. Schellekens, "Boson Mass Spectrum Of Kaluza-Klein Theories On Hyperspheres," Phys. Lett. B **143** (1984) 121.

[102] H. Georgi, "Lie Algebras In Particle Physics. From Isospin To Unified Theories,", second edition Front. Phys. (1999).

[103] E. T. Whittaker and G. N. Watson "A course for Modern Analysis," 4th Ed. Cambridge, England: Cambridge University Press, 1990.

About the author

Alberto Salvio is a visiting professor at the institute of theoretical physics at the Autonomous University of Madrid. He obtained his Ph.D. degree from the International School for Advanced Studies (SISSA) of Trieste in 2006 under the supervision of Seif Randjbar-Daemi.

Alberto Salvio during a workshop in Cambridge

His main research interest is in the physics of fundamental interactions and he taught general physics courses as well as theoretical physics courses at the Ecole Polytechnique Fédérale de Lausanne and the Autonomous University of Madrid.

www.ingramcontent.com/pod-product-compliance
Lightning Source LLC
Chambersburg PA
CBHW072351200326
41519CB00015B/3734